LIVING ON THE MARGINS: SOCIAL ACCESS TO SHELTER IN URBAN SOUTH AS

T0231200

Living on the Margins: Social Access to Shelter in Urban South Asia

NAVTEJ K. PUREWAL

Development Studies/Sociology, University of Manchester

Routledge
Taylor & Francis Group

LONDON AND NEW YORK

First published 2000 by Ashgate Publishing

Reissued 2018 by Routledge
2 Park Square, Milton Park, Abingdon, Oxon OX14 4RN
711 Third Avenue, New York, NY 10017, USA

Routledge is an imprint of the Taylor & Francis Group, an informa business

Copyright © Navtej K. Purewal 2000

Notice:
Product or corporate names may be trademarks or registered trademarks, and are used only for identification and explanation without intent to infringe.

Publisher's Note
The publisher has gone to great lengths to ensure the quality of this reprint but points out that some imperfections in the original copies may be apparent.

Disclaimer
The publisher has made every effort to trace copyright holders and welcomes correspondence from those they have been unable to contact.

A Library of Congress record exists under LC control number: 00109565

ISBN 13: 978-1-138-72877-6 (hbk)
ISBN 13: 978-1-138-72871-4 (pbk)
ISBN 13: 978-1-315-19040-2 (ebk)

Contents

List of Figures

List of Tables

Acknowledgements

There are a number of individuals and families to whom I am greatly indebted for their roles in the process and completion of this book. First and foremost, to the two closest to me, Virinder for being there every step of the way and Nuvpreet for inspiration and meaning. To my own parents and brother for their good wishes along the way. To my parents in-law for their unconditional moral and practical support. Thanks also go to the extended Kalra, Purewal, Sira and Bedi families for making the fieldwork experience in Punjab like being at home.

For hospitality and institutional sponsorship at Guru Nanak Dev University (Amritsar), I am grateful to Dr. Jasmeet Kaur Sandhu, Professor Ranvinder Singh Sandhu and, of course, Sandeep. A special thanks to Dr. Harjinder Singh for insightful advice and direction and also to Professor Harish Puri. To Pankaj Bawa and Harpreet Ghai for showing me the streets of Amritsar on the back of a scooter and whose diligence in conducting the survey are greatly appreciated. To the Mann family for their kindness during the year in Amritsar. For helping to make a home away from home, to Asha.

On to my support in Britain, to Laura, Kath, Kathy, Sanjay, John and Raminder for friendship and encouragement. A special thanks to Nilofer, Kabir, Shahina, Bilu and Manjit for their patience and understanding. In the U.S. and beyond, to dear and old friends Mamta, Jenn and Kim. To the Punjab Research Group for continuing to pave the way for new and much needed research on Punjab. To Dr. Tasleem Shakur for supervision and guidance on the overall study.

Finally and most importantly, to the families living around the walled city in Amritsar whose struggle for decent living conditions continues.

Introduction

The Focus

The opportunities available to the urban poor in the Third World to find decent accommodation for themselves are becoming increasingly narrow. Those without the monetary savings, steady incomes or the privilege of contacts required to afford or access adequate shelter find themselves on the margins. In actuality, the familiar and often stereotyped squatter settlements and other extra-legal, informal types of settlements provide a considerable proportion of the housing supply to the poor. Meanwhile, in many Third World countries the state has historically not played a particularly magnanimous role in providing social welfare. Furthermore, the recent 'down-sizing' of the state's role in social welfare provision has already had profound impacts upon the patterns of access to shelter that are now emerging. Privatisation policies, as will be argued in the book, largely rely upon already existing 'self-help' settlements which continue to attract the poor and to be the largest housing providers in many cities. As part of the same process, existing social inequalities have been exacerbated in poor residential areas without the full commitment of official policies to strive for social equity. The withdrawal of the state apparatus from direct housing provision has left many households to their own devices to find shelter in an increasingly commercialised, unaffordable sector while the parallel phenomenon of the state's encouragement of private activities has further entwined the low-income housing sector with market forces. The emerging patterns of social access to and exclusion from shelter show a socially differentiated picture which requires an interrogation of the appropriateness of certain policies and the reliance upon individual household efforts.

The poor residential communities surrounding the walled city in Amritsar are the focus of this study. They occupy a geographical place which is on the margins of the walled city as well as a metaphorically marginal position within the housing market of the city. In developing an understanding of the dynamics of development in the low-income settlements in Amritsar, the study offers a framework for examining the experiences of low-income settlements specific to the city. This book

1

highlights a number of theoretical points. The first is that the self-help debate still holds many points of relevance to contemporary housing research. While the role that the state has played in development, and more particularly housing development, has evolved over the past few decades, the ideological debates around the extent of the state's capacity in mobilising resources for social welfare and that of the market and individual's role in the development process still remain open to discussion. Reforms to state policies from provision to assistance have fundamentally impacted upon the emerging types of settlements which often cross and challenge the dichotomy of categories like legal-illegal, formal-informal and public-private.

In developing a framework for assessing social access to housing, a comparison of some of the competing debates, mainly between the neo-marxist and self-help theorists, is made here. This is complemented by reference to recent comparative empirical studies on the formation and consolidation of low-income settlements. In doing so, the book attempts to assess the ways in which self-help and planned intervention have impacted on low-income housing systems. Patterns of social differentiation created and reinforced by both forms will be comparatively analysed in the final analysis as a means of drawing together the core issues around social access.

The Organisation of the Book

The first chapter reviews several of the themes in the literature on poverty and shelter in the Third World. These themes are addressed through the main competing theoretical perspectives within the literature of the self-help school and the neo-marxists. The ascension of self-help thought starting from the 1960s saw the ideological split between those who saw potentials in the efforts of individual households in housing improvement and those who saw the use of self-help methods as exploitative of the poor. This divergence is reflected in the case presented by the self-help theorists in favour of autonomous housing systems and the neo-marxist critique of the use of petty commodity production under capitalist conditions. The theoretical thrust of this study is introduced here in the context of the neo-marxist critique, which is further elaborated upon in the next chapter.

Chapter Two traces the arguments around housing access both historically and ideologically through the implementation of policies. The effects of specific policies such as legalisation, upgrading and conventional programmes upon social access are addressed in relation to the theoretical debates of Chapter One. Here self-help and its critique are analysed in practice. Trends within housing policies in the wider international context are then examined through a look at conventional and non-conventional methods. This considers the broader spectrum of self-help and state-assisted means of housing supply through a continuum of unaided self-help, state-supported self-help, state-initiated self-help and conventional housing policies. In breaking with the tendencies in both policies and the literature to use fixed criteria, Harms' (1992) continuum is adopted as a useful method for providing a framework for understanding the diversity of types of housing being accessed by the poor.

The next section is concerned with the interest within the literature on illegality. The nature and consequences of settlement illegality are diverse, though overlapping, and are explored through theoretical and empirical evidence from studies of different Third World urban contexts. However, the official views upon illegality, as will be argued, have not significantly changed over the past few decades with the dichotomy between legality and illegality predominantly remaining intact. The final section of this chapter presents the shift towards research on unregulated housing sub-markets as a more reflective and contemporary analysis of the rapidly evolving nature of Third World housing. The variety of methods by which the poor gain access to housing challenges the legal-illegal dichotomy that had traditionally been upheld. Chapter Two concludes with examples of studies on a number of cities which further highlight the increasing relevance of the unregulated housing sub-markets in Third World housing studies.

Chapter Three offers the contextual background to the study with a brief discussion of history and social change in Amritsar. The region of Punjab has experienced a number of dramatic changes which have affected the demographic, social and political conditions of the region. The development of the city is traced from its inception to its expansion through the *katras* and the rise of the kingdom of Maharaja Ranjit Singh after which the period of decline commenced. British colonisation in 1849 and the eventual partition of Punjab in 1947 had lasting effects upon Amritsar's position as a major urban centre. The chapter also concludes with a brief overview of some of the major events affecting ethnic and communal politics in the region, including the 1966 linguistic reorganisation of states and the storming of the Golden Temple in 1984. The history of decline of the city of Amritsar is paralleled to the structurally marginal position of the city's poor, low-caste population. The themes addressed in this chapter re-emerge in subsequent parts of the book as the position of poor communities in the city has been marked by historical processes of economic development, political repression and social segregation.

In Chapters Four and Five I present the methods used to develop my theoretical approach in the fieldwork. This involved rectifying the various terms and terminologies which have been used to discuss poverty and housing issues. A central part of this methodology was the construction of a typology which is designed specifically for the context of Amritsar according to social, developmental and physical characteristics of settlements. National and state-level housing policies in India and Punjab are touched upon where similar trends to other Third World country experiences, as noted in Chapter Two, are identified. This puts into perspective the policy backdrop of public sector involvement in housing provision. On a more practical level, the selection of the sample and the process of carrying out a survey are briefly overviewed with a breakdown of the sample and spatial identification of settlement locations. The typology, as introduced here, reappears throughout the remaining empirical chapters as it forms the basis for analysis of social access.

Chapter Six introduces a number of social categories which have significance in the socio-economic make-up of poor communities in Amritsar. Caste and religious identities are represented in the sample as distinct groupings which, while having experienced a certain amount of change through the historical development of the region, still have a place in the demarcation of communities and the settlements in which they live. The chapter goes on to analyse the position of migrants and the diversity of migrant communities in the city. The experience of migration and the impetus to migrate is by no means singular, and the case of Amritsar exemplifies this. The position of migrants in low-income settlements is revisited in Chapter Eight where Turner's theory of migrants and mobility is critically analysed. Economic conditions such as income and employment opportunities also form an important part of the socio-economic profile of the sample, as certain occupations have traditionally been associated with specific caste communities. Income and employment also have implications upon where various income groupings fit into the housing market.

Chapter Seven concerns itself with processes of settlement. The ways in which the survey settlements have developed and consolidated are considered with regard to the choices available to households, the location of housing and the means used to acquire houses. This is done to illustrate the operation of the housing system as introduced in Chapter Five and also to identify predominant forms of settlement within the three-tier typology. This is followed by an examination of housing and land tenure in which tenancy, land ownership and the effects of legalisation upon tenure patterns are included. The state's legalisation policies of squatter settlements on the one hand has resulted in increased access while on the other hand private development has put increased pressure upon scarce public land resulting in oftentimes further 'illegalisation' of the poorest settlements. The chapter concludes with a brief analysis of the relationship between security of tenure and households' expressed intentions of remaining in their current homes.

Chapter Eight is a culmination of the previous chapters in its empirical focus. The first section sketches the paths that are available to poor households in obtaining housing through a schematic diagram. The diagram exhibits the flows of social groups within the typology and the routes available to new migrants, local urban poor and old migrants in the housing structure of Amritsar. The next section examines the use of social status as a measurement of housing access initially through Turner's model of migrant categories and subsequently through evidence from the survey in this study. A number of different variables, previously introduced in Chapters Six and Seven, are then applied to the typology as a means of assessing certain aspects of the social dimensions of access. The chapter concludes by presenting a picture of the social dimensions of each type of housing as a way of drawing conclusions about patterns of access in Amritsar's low-income settlements.

The final section of the book draws some conclusions about emerging patterns of social access to housing in various local and national contexts, drawing largely upon this study of Amritsar's low-income settlements. The inter-linkages and gaps between the self-help and public sector housing supply systems has shown a highly diversified and complex picture of shelter arrangements being accessed by the poor. Social differentiation within this complex system, from the evidence of this study, is argued to be accentuated by the nature of the state's interventionist though non-redistributive approach towards housing and land for the poor. Detailed histories of settlements surveyed in the study can be found in Appendix 2.

1 Poverty and Shelter in the Third World

Introduction

The presence and growth of slums and squatter settlements in so many cities in the Third World has generated much debate in the literature on poverty and shelter in the Third World which has primarily been concerned with explaining the processes and conditions in any given society that create them. There has been a distinct polarisation of analyses in this respect. Market analysis, on the one hand, has accounted the shelter problem of the poor to a gap between supply and demand. Market approaches to the shelter question have inspired the self-help school and a variety of policy schemes to improve and legalise many settlements through the aid of international development agencies. The political economy analysts, meanwhile, have attributed the inadequate housing situation of the poor to structural functions and requirements of capitalism rather than a shortage of supply. Political economy and neo-marxism have been the main opponents to the prevailing market analysis and self-help approach. This chapter will overview the views on poverty and shelter of these two schools while also introducing the theoretical thrust of this study which will be further developed in Chapter Two.

Theoretical Perspectives

The growth of slums and squatter settlements in cities of the Third World, according to market approach analysis, is a result of rapidly increasing population growths to which governments are unable to provide basic infrastructure, housing and services (Payne 1977; Palmer and Patton 1988). To deal with the mounting low-income housing problem requires a policy to combat the so-called housing shortage. This has been the most dominant approach in the formulation and implementation of housing policy and has been widely championed by the World Bank, the UNDP and national government schemes. As a result the self-help philosophy has been adopted

6

with vigour through sites and services and upgrading schemes which were intended to decrease the gap between the supply of houses and the increasing population figures (van der Linden 1986).

With low-income housing problems in Third World cities viewed as a housing deficit situation, the remedies offered by the World Bank have been ones which have sought to strengthen the supply side to meet the demand for housing (van der Linden 1986; Pugh 1995). In order to do so, the supply costs must be brought down so that access to official or legal means for housing and services would be made available to more people (van der Linden 1986). Land, supply and delivery of services, and finance have been constrained by institutions because of prices and standards out of reach of the urban poor. The market approach argues that while standards and costs should be brought down in order to ensure delivery to the poor, people should be encouraged to build their own houses in order to reduce the usage of public resources. Thus, economic viability *vis-a-vis* cost recovery analysis is used as a criteria for the justification of most World Bank-sponsored programmes (Moser 1982).

Within this framework the formation of slums and squatter settlements is viewed as a process of people settling where there is demand for labour. Slum formation is therefore seen as a natural process by which the poor respond to market forces which offer opportunities for them to better their lives. This approach argues that as residents of slum and squatter areas are active participants in the local economy, it is counter-productive to see them as a negative element (Payne 1984). Instead, efforts should be made to improve their living situation in order to facilitate their contributions to the economy. In a study of affordability and planning of housing in India, Mehta and Mehta (1994) argue that there is need for a market-based perspective on affordable housing and that only the market can bridge the gap between housing demand and supply. They render the "conventional approaches of (the) welfare state fulfilling basic needs a limiting factor (Mehta and Mehta 1994: 20)." This approach has had strong effects on governments who perhaps had once administered clearance programmes, but, who now acknowledge and have even legalised and upgraded certain settlements so that the availability of affordable serviced land for low-income households can reach the demand (Mehta and Mehta 1994; Payne 1984).

The competing theoretical perspective of political economy places issues of poverty and housing within a larger world political economy framework. The 1970s saw new theoretical perceptions of urbanisation processes through the works of Frank (1967), Amin (1974) and Wallerstein

(1979). This approach critically examines world capitalism and the international economic system by focusing on the role that Third World nations and, more specifically, cities play in the accumulation of capital and its subsequent filtering through to developing countries. The importance of land, capital and labour has shaped the organisation of the city in such a way as to facilitate production as well as to operate according to the needs of those controlling the means of production (Harvey 1989; Castells 1978). This framework sees Third World cities as playing a central role in the underdevelopment of the Third World and the development of the developed world (McGee 1985).[1]

The low-income housing theories of the 1950s and 1960s have been strongly criticised by the political economy theorists who argue that population, housing shortage, and lack of government funds are not the root of the housing crisis but that it is incumbent structural issues which need to be radically altered. In this light, the self-help arguments and subsequent planning methods are merely extensions or temporary remedies to conditions created by capitalist modes of production. These seek to push the responsibilities of basic needs such as shelter away from the state and the collective and onto the individual (Burgess 1984).

A major theme in the political economy approach has been its acknowledgement of the interaction between the capitalist and peasant modes of production.[2] In the realm of low-income housing this has had profound effects since a large proportion of slum and squatter housing is supported by the informal sector, and the employment and housing options available to the poor are increasingly being delivered through informal networks. Within this framework the articulation between capitalist and peasant modes of production is argued to be inherently exploitative of the poor. The interests of capitalists are served by the growing pool of poor labour who migrate to wherever employment opportunities can be found. The result has been the considerable expansion of the informal sector which has become increasingly important both to the sustenance of the formal sector and to the livelihoods of the poor. However, the informal and formal sectors are increasingly becoming inter-related as both are dependent on one another. The existence of the informal sector is profitable for the formal sector as it provides cheap labour and materials, and the informal sector is generated by the wide-ranging economic activities of the formal sector.

Along with income generation, housing for the urban poor is provided by the informal sector in terms of construction, maintenance, services, and land. The increasing marginality of the urban poor from legal frameworks has meant that employment, housing and basic needs are met through the most accessible means as possible. The informal sector provides these needs to a vast majority of the poor in Third World cities and will continue to do so as long as no other infrastructure exists (Hardoy and Satterthwaite 1989). Particularly in the absence of welfare state principles in developing countries, basic housing for the urban poor has been made a luxury and not a right. The role of the poor in free market economies is to provide their cheap labour for the further advancement of capital accumulation. Even small-scale, pre-capitalist production is transformed into capitalist production, virtually eradicating all forms of small-scale, self-help production (Burgess 1984). The implications on housing are that people no longer have to contribute to the building of their houses but instead rely upon forms of housing production which are part of the dominant capitalist mode of production and as a result increase the value of capital. Housing is therefore no longer simply a basic need through use-value but has also become a commodity through exchange value, the value of which can only be realised by those who have a housing need and who can afford to buy it.

The frameworks presented by the two main schools of thought on slum and squatter formation have been briefly summarised here. The market analysis theorists have understood the emergence of slums as the result of the gap between the supply and demand for housing while the political economists have responded to the structural conditions which they see as generating poverty and the housing shortage resulting in the proliferation of slums. The political economy approach focuses upon the structural factors governing production and distribution which result in the marginalisation of the poor. The following section will focus more specifically upon how each of these perspectives relates to debates on poverty and shelter, namely through the market approach-inspired self-help school and the political economy approach-inspired neo-marxist critique.

Self-help

As mentioned in the preceding section, there is a parallel between the political economy analysis within wider development theory and its application to poverty and housing. Similarly, the alignment of the self-help school with the prevalent neo-classical and neo-liberal models of development also follows a similar path. The self-help approach to housing for the urban poor has been the dominant discourse in both the housing literature and policy formation in the past few decades. Through the sites and services and upgrading projects of international development agencies, the self-help approach has been taken to heights which have extended the concept contrastingly from the inception of the term in the 1960s and 1970s (van der Linden 1986). Self-help individual utility maximisation and empirical market analysis have been brought together in the institutional context of self-help housing. The structuralists, here referred to as neo-marxists, have responded to the orthodoxy of self-help with theoretical attacks upon the effects of individual solutions upon the household economy as well as upon wider social, political and economic structures.

Nothing exhibits this ideological debate more succinctly than the dialogue between John F.C. Turner and Rod Burgess in the late 1970s (Burgess 1978; Ward 1982). In this dialogue Turner defends his support of self-help as an apolitical one (Turner 1982). Burgess, meanwhile, sees the basic characteristic of housing within capitalist systems to be a commodity and therefore open to market forces which inevitably work to the disadvantage of the poor. The basic analysis by the self-help school begins by emphasising housing as a social necessity, and, with a diversity of economic, social and cultural circumstances, people are themselves the best providers of housing. Market analysis of the housing situation, through the dominance of the self-help school in policy implementation, has placed the emphasis upon the gap between the supply and demand for housing, thus arguing that the individual's role is to provide the housing while the state's role is that of a facilitator (Marcussen 1990). While some would argue that the phrase 'self-help' rids itself of any political affiliations (Turner 1982: 99), the overall debate has shown a distinct division between those who are critical of capitalist development and see the need for a more comprehensive state in social welfare provision and those who see a the role of the state as a facilitating, though minimal, one. This reflects the highly ideological nature of self-help analysis and its critique with the role of the state and the individual central to the competing arguments.

The concept of self-help only came to be defined through the theoretical upheaval of the late 1960s and early 1970s when a number of authors presented their individual interpretations. Charles Abrams used the idea of auto-construction[3] as his definition of self-help. However, he was wary of the use of auto-construction as a solution to housing inadequacy in cities. His idea of self-help is represented in *Housing in the Modern World* (1964) in which he presents the concept of core housing where basic infrastructure and wet components would be provided by the state while the further construction should be done by owners through step-by-step processes. This, Abrams argues, can be achieved through a housing policy which encourages the development of financial institutions and savings on the part of individual families. He goes on to conclude that home ownership is one of the prime hopes of poor families in the Third World as ownership provides lifetime security and that "the market place and the widespread ownership of land and shelter are still the seeds from which larger freedoms have grown and can continue to grow..." (Abrams 1966: 296).

Almost simultaneous to Abrams' work, John Turner, an architect, and William Mangin, a sociologist, observed the slum situation in the 1960s in the *barriadas* of Lima, Peru where no other solution than clearance had previously been considered by government authorities (Marcussen 1990). Turner and Mangin, through their own investigations, found that the slum communities were highly organised and economically dynamic areas. They also discovered that there existed a housing system which was based on gradual improvement and incremental construction which correlated with the changing economic situations of individual families (Turner 1968). Out of this 'discovery' emerged ideas of self-help which were based upon the power of individual efforts and potentials that lie in people's own abilities to help themselves.

Turner's observations were based upon what he had seen in autonomous settlements, whether informal or illegal, in which the dwellers controlled the major parts of the housing process (van der Linden 1986). From this aspect he developed the concept of user control in which the focus was on decision-making rather than on labouring (Turner 1976). The importance of incremental building under user-control outlined Turner's attitudes and can be summed up in the well-known phrase: "housing as a verb rather than a noun"[4] (Turner 1978: 1137). His emphasis on local communities and user-control within an autonomous system put Turner's theory in direct conflict with the heteronomous systems of socialist economies which rely upon centralised decision-making bodies and large-scale technologies (van der Linden 1986).[5] At the same time, however, he argues that in order to make legal housing more affordable to the poor of Third World countries, the quality of standards should be made lower (Turner 1976). This, as will be later discussed, is in direct opposition with the neo-marxist critique which rejects moves towards reducing standards.

According to Turner, the political debates that have arisen out of the self-help discourse are trivial to the actual potential that self-help holds and do not represent the effectiveness of individual and small group activity in making their own decisions about their own housing (Turner 1978). Instead, he sees the debate as being between "those who assume, consciously or unconsciously, that material economy depends on large-scale production and supply systems, and those who do not" (Turner 1982). He sides with those who see the limitations within centralised housing supply systems and who contest that the only ones able to efficiently and properly administer the resources for housing are the people themselves. However, he works on the assumption that housing should not be centrally administered. The units and resources provided would be assumed to be standardised and Turner argues that rigid systems would be applied to extremely variable demands and needs (Turner 1976). Therefore, inappropriate procedures and schemes would be applied to situations with different dwelling and neighbourhood types and locations, different forms of tenure, and individual household practices. The alienation of the individual from planning and administrative processes would result in what Turner refers to as a 'mismatch,' affecting the individual's desire to invest in and care for the property (Turner 1982). Centralisation would create more of an obstacle than it would facilitate in the improvement of housing conditions of the poor.

A perhaps less polarised though sympathetic position on self-help is that of Moser (1987) who stresses the importance of developing a feminist perspective to the Third World housing debate. While she is critical of Turner's version of self-help housing, she only goes so far as to assert that it is "through women's participation and mobilisation can challenges to the sexual division of labour be made" and that only "through the examination of the relationship between gender and housing...(can there be a) ... debate on the importance of housing and human settlements not only as an economic relation but also as a social relation" (Moser 1987: 5). Moser welcomes the move towards sites and services and upgrading policies as a shift in approach from top-down to more diverse self-help solutions which hold potentials for the active participation and inclusion of women in the housing process.

A number of other studies emerged in the 1980s which reflected upon the experience of self-help housing in its policy form. Skinner and Rodell's (1983) edited volume on the problems of self-help housing[6] noted the experiences of a variety of different self-help programmes as evidence of the vast potentials of self-help in its "infinite number of possible combinations between governmental and family investment in housing" (Skinner and Rodell 1983: 2). Payne (1984) brought together a number of studies in another edited collection attempting to specifically analyse the theoretical and practical implications of sites and services and upgrading. While the achievements and limitations of the policies as implemented in the various different Third World countries are reviewed, a recurrent theme throughout the book is that the immediate results of self-help policies are an advantage over the bureaucracy of conventional systems through their centralised planning and decision-making, large-scale production and time delays in delivery.

Turner's theories of self-help have thus become the basis for the institutional context of self-help housing as seen through sites and services and upgrading.[7] Turner had continually emphasised his position on autonomous housing systems against state-oriented, centralist approaches to justify his argument for self-help. Despite his claims of the apolitical nature of his definition of self-help, he essentially enters the debate between the political left and right. Similarly, Turner, who prefers to separate himself from the bulk of housing experts, has come to be known less as a radical than as the founder of the current consensus, the Intermediate Technology school, on housing in Third World cities (Burgess 1982). The subsequent emergence of self-help oriented projects and a body of literature on the position of self-help approaches in housing policy has further polarised the self-help school from its critics who now not only have theoretical arguments but also the experience of the effects of self-help policy through projects implemented in many different Third World cities.

The Neo-Marxist Critique

Marxist theory first and foremost identifies housing under the capitalist mode of production as a commodity. Housing, in this sense, is produced, exchanged and consumed in a manner which is determined by production (Marcussen 1990). Rod Burgess and the other neo-marxist theorists apply this analysis to the Third World context arguing that housing and land cannot be thoroughly understood through their use values alone but must be investigated through processes of economic production and social reproduction.

Further elements of the 'self-help' debate involve the distinction between market-value and use-value which both Turner and Burgess include in their works. Burgess focuses on the capitalist mode of production and class analysis in his approach to housing. He looks first to the relationship between the dominant capitalist mode of production which produces industrialised housing products. The relationship between the production and consumption of the industrialised housing product is determined by market exchange. He also considers the dependent petty-commodity form of production which produces self-help forms and manufactured forms, where there is less of a distinction between market and use value (Burgess 1982). In the dominant industrial mode the relationship between the agents involved in the housing sector and the housing product are set according to the market as opposed to through patron-client relations (Burgess 1982). The industrialised form of production of housing materials monopolises the production of building materials such as cement, brick, and roofing. It controls the supply of land and it allocates housing budgets while the petty-commodity form of production bridges the gap between the market-value and use-value by creating less dependence upon the market and more dependence upon need and consumption.

The petty-commodity mode of production of housing takes on two forms: the manufactured and the self-built. The manufactured is produced primarily for its exchange-value and the self-built for its use-value. In the self-built form there is no distinction between the producer and the consumer. Generally, the materials used are recycled and the labour inputs are either familial or free. This social system, though functioning under capitalist conditions, resembles a pre-capitalist setting, one which relies heavily upon kinship work traditions and ethics. However, with regard to both land and physical inputs the conditions have strong connections with capitalist structures in that private land is a commodity and the material inputs are either market-valued or are market-waste. So while self-improvement efforts may represent pre-capitalist social structures, the existing economic conditions are capitalist ones which therefore hold traditional economic activities as unimportant or unvaluable.

In his analysis of self-help efforts within pre-capitalist traditions and capitalist structures Burgess labels Turner's approach as draconian and particularly criticises him for depoliticising the housing issue (Burgess 1982: 74). Burgess questions whether Turner sees the housing problem in urban areas as a crisis in the capitalist mode of production or even as a structural condition of capitalism since he does not address the issue of structural changes in the political process nor does his political analysis go past the ineffectiveness of state intervention in housing and land markets. It is clear that Turner does not acknowledge the potential of class struggle over access to housing and services. The dynamic potential within self-help housing improvement to Turner is the petty-commodity mode of production. Burgess is critical of this as being Turner's main means of meeting housing needs. Turner, focusing on more technical aspects of housing which he proposes could be met through petty-commodity production, does not address the issue of access to resources which is of a fundamentally economic and political nature.

Burgess is also critical of Turner's ideas in terms of the possibility of expansion of self-help programs. Firstly, he argues that the prices of basic building materials would increase while recycled materials would also be given new market-value as well as a new use-value with the likelihood that private interests would get involved. Secondly, he points out that state-assisted self-help programmes would merely be a response to the arising crisis in capitalism and hence the crisis in housing. With the rise to dominance of capitalism, processes of production have been converted from use-value to commodity production based on exchange in the market. Capital accumulation has become the driving factor in production while production for necessity has become virtually obsolete. Self-help in this sense could be called pre-capitalist since there is a close relationship between the use and life of the people who build the houses (Harms 1982). It is this combination of use-value and exchange-value that Burgess' critique (1978) of Turner is based upon and which forms the foundation of the central arguments in favour of the efficiency and 'common sense' of self-help housing policies.

The ideological implications of the debates around self-help have continued to be a cause for disagreement between the self-help and neo-marxist schools.[8] With increasingly expensive markets for housing, according to self-help, people themselves can initiate self-help methods individually or in groups in order to cope with the unaffordable costs of adequate housing. Burgess argues that the practice of people constructing their own homes through self-help is an age-old practice. State supported self-help only dates back to the nineteenth century in Europe when Burgess (1992) notes that state-sponsored self-help arose out of the failure in modernisation policies to trickle down the benefits of growth to the urban poor. With the realisation that modernisation was not producing the expected results, the bulldozing of slums and squatter settlements was virtually abandoned in many Latin American, African and Asian cities as there were few positive impacts on housing needs, budgets and public opinion. In his reference to the activities of the Alliance for Progress in Latin America which is widely recognised as an extension of United States foreign policy in the region, he goes on to describe the transition from conventional housing policies to non-conventional policies by many Latin

American, Asian and African countries.[9]

Through the 1970s and 1980s economic development policies in the Third World encouraged basic needs and redistribution with growth, which resulted in the international sponsorship of self-help housing projects (Burgess 1992). Policy recommendations made by Turner and Abrams were interpreted as highly compatible with economic policy at that time and became the foundation of World Bank housing policies after 1974 (Moser 1987). The extent to which the institutional capacity of the World Bank and other international agencies in providing replicable and appropriate solutions to problems of housing and infrastructure in Third World cities has been discussed by a number of researchers (Rakodi 1991; Payne 1984; van der Linden 1986; Miah et al 1988). The exploration of case studies from Latin American, African and Asian countries in assessing the experience of sites and services and upgrading has extended the body of research on institutionalised self-help projects, which will be highlighted in the next chapter.

The self-help approach has, while claiming not to have political affiliations, been contested for detracting attention away from the rights of individuals to have proper housing and towards the ability of people to help themselves in its argument that 'people know best'.[10] Through the attacks by the self-help school upon conventional policies, the public sector has been aligned with inefficient top-down approaches while the individual, and more recently the private sector, has been identified as the more efficient means to addressing the housing situation. This shifting of the burden away from the public sector onto the private sector has meant that the overall responsibility has been left with the poor to pay and plan for their own housing, although the facilitating role of the public sector has been made explicit in self-help policies. As a result there has been an increase in the role of the informal economy in many developing countries as the major provider of housing to the poor while state assistance in collaboration with self-help housing provision has proven to be less than adequate (Skinner and Rodell 1983).

The way in which the informal sector has filled the gap between the needs of the poor and the state has added another dimension to the theoretical debate between the advocates and critics of the self-help school. The informal sector has emerged, to a large extent, as a result of the already-existing self-help activities of the poor which remain unchecked by official institutions. The state's exoneration of its role in directly providing housing has meant that the demand for housing and other basic needs have been predominantly met by self-help efforts. Meanwhile, the shift in policy has shown a dramatic movement towards the role of the state as a facilitator of self-help efforts rather than as that of an initiating force in housing supply. The effects that this shift has had upon access to housing among the urban poor has led to the demise of conventional housing schemes and in the upsurge of sites and services, upgrading and land policies which have produced diverse and socially differentiating results. Elements of this trend are evident in the empirical study of this book where informal means of accessing services and information through local political and economic interests contribute to the paving of paths for the poor to gain access to shelter.

While wider debates within the social sciences have shifted since the political moment at which the self-help/neo-marxist confrontation took place, the theoretical framework adopted in this study maintains the importance of the core arguments between these two schools. I would argue that the theoretical concerns with capitalist development and the role of the state in housing provision are still as pertinent as they were over two decades ago. In fact, the luxury of hindsight of the collusion of self-help with capitalist modes of development provides a particularly strong base from which to continue the critique of self-help theory and practice. While divorcing itself from other dichotomous paradigms which are now dated in their simplistic understandings of legal and formal frameworks, this study will attempt to modify the neo-marxist critique of self-help. By applying its theoretical and practical concerns to the primary and secondary empirical research, this book will question the underlying assumptions of self-help while also analysing some of the more discreet social outcomes of self-help policies in conjunction with the devolving role of the developmental state in this area. Chapter Two will clarify this theoretical position by further delving into the more specific debates on housing access and setting the stage for empirical chapters to follow.

---

[1] Wallerstein (1976:30) argues that the global economy has been formed by both the expansion of Europe and through the expansion of capitalist forms of production.
[2] See International Labour Organisation (ILO) (1972), *Employment, Incomes and Equity: A Strategy for Increasing Productive Employment in Kenya*, ILO: Geneva.
[3] Auto-construction in this sense literally means 'people building their own homes with their own hands'.
[4] Turner in this phrase states that housing should be viewed as a process and activity rather than as a product.
[5] Though belonging to the Intermediate Technology School, Turner (1976) also points out that a varying range of technologies should be implemented in order to meet appropriate housing needs. This also includes centrally administered

technologies which leads Turner to not totally discard the role of the state or central authority.

[6] See R.J. Skinner and M.J. Rodell (eds.) (1983) *People, Poverty and Shelter: Problems of Self-help Housing in the Third World*, Methuen: London.

[7] In Chapter Two the different interpretations of self-help through institutional support will be more extensively discussed.

[8] A parallel debate around political ideology took place between Manuel Castells and the urban sociology school, namely the Chicago School (Dickens 1990). The self-help debate and the positions of the public and private spheres in housing production can be positioned within Castells' analysis of the role that the state plays in capitalist societies where infrastructure such as housing are no longer provided by the state but are instead mediated through state-supported self-help policies.

[9] In Chapter Two the positions on housing access will be reviewed in the context of the historical developments in housing theory and practice in Third World countries.

[10] See Turner, J.F.C. (1976) *Housing by People: Towards Autonomy in Building Environments*, Marion Boyars: London.

2   Access to Shelter:
Issues, Debates and Policies

Introduction

The rise to dominance of self-help can be contextualised within the historical background of the demise of welfare principles in the West and its imposition upon policies adopted in developing countries at the national and international levels. It is for this reason that the self-help philosophy became popular starting from the 1960s to the present. While policies such as legalisation and upgrading of squatter settlements were part of the initial wave of self-help, the outcome of many such efforts was plagued by overriding market forces which in some cases resulted in the gentrification of areas which became inaccessible to the poor. Similarly, in the case of many conventional housing projects, there has been evidence that middle and upper middle income families have been occupying social housing meant for lower income groups where the competitive housing market has squeezed them out. Therefore, it is quite clear that the picture of the shelter situation of the urban poor is continually evolving and cannot be simply reduced to stereotypical squatter settlements and slums. In highlighting some of the key issues surrounding access, this chapter will further examine some of the experiences of various policies upon access and will conclude by eluding to a framework for understanding poverty and shelter outside of the slums/squatters stereotype.

Ideological Discourses on Access

Whether the state's support of self-help facilitates wider access or if it further accentuates social inequalities has been addressed from both sides of the low-income housing debate. The effects of institutional support of self-help upon access has been a particular point of contention. Burgess argues that government-sponsored self-help policies are:

> proposals for the maintenance of the capitalist mode of production...

(and)... representing the interests of the various fractions of capital tied to housing and urban development ...(which) reflect the interests of the dominant fraction or fractions (Burgess 1978: 1126-1127).

Meanwhile, self-help advocates have conversely argued that self-help policies increase access to a wider pool of households. Rodell and Skinner view the institutionalisation of self-help as a means for the government to supplement the 'missing elements' while reducing their expenditures:

> By combining the housing investments of families and governments, governments might reduce their investment per family and so reach a larger number of families, thus helping to overcome the main deficiency - low access - which resulted from conventional housing (Rodell and Skinner 1983: 1-2).

The main ideological push in the direction of state-sponsored self-help housing occurred during the late 1950s and 1960s when social welfare and urban development were becoming of concern to Third World national governments and international development agencies. The logistical implications of providing housing to the large numbers of people requiring shelter made the 'spontaneous settlements' of Latin America an exemplary case of how urban housing could be addressed through greater 'user control' (Harms 1982; Alexander 1988). During the post-war period in Europe, multi-storey prefabricated systems had been constructed as a means of quantitatively fulfilling the housing deficit. Simultaneous to the criticism of the experiences of public housing projects in Europe, comprehensive housing schemes were being undermined by developing countries and development agencies during the 1950s and 1960s. The post-colonial Third World states which had emerged from the middle of the century were in a crisis of resource allocation and nation-building to which social welfare became an agenda for economic planning. For example in India the Planning Commission introduced Five-Year Plans in 1951 to consolidate its development goals under which social welfare was one of the themes.

Burgess (1992) views the rise of self-help policies within the ascension of modernisation and redistribution with growth (RWG) strategies in the post World War II period, despite the growing popularity of dependency thought through such writers as Frank (1969) and Rodney (1976). The modernisation approach requires the growth of industrial development, in the case of housing, through the construction of industrialised modern houses based on Western technical and cultural standards (Burgess 1992). However, Burgess cites the expansion of the modernisation approach in the example of the Alliance for Progress in Latin America as having exhibited the political objectives of both spreading the ideology of modernisation while also appeasing the masses in light of the threat of replication of the Cuban Revolution to other parts of Central and South America. The highly politicised nature of self-help housing policies has been further examined by a number of other researchers (Gilbert and Ward 1982; Mathey 1990 and 1992; Ramirez et al 1992) by whom the links between widespread poverty and housing with political patronage and control have been made explicit.

At the forefront of debates on access have been justifications for and against conventional housing projects and institutional self-help policies. There is a distinct departure between the critics and supporters of institutional self-help with regard to access in terms of the relationship between labour, costs and standards. Conventional housing projects have been the object of much criticism during the past two decades during the period when self-help policies had ascended on an international level. The case against conventional policies has been most commonly made on the basis of high costs, insufficient targeting and inefficient use of resources. Rodell and Skinner (1983) state that conventional housing's failure was due to an inability to reach the low-income families for whom they were targeted and instead that state-supported self-help "might improve chances of the poor to benefit from government housing investment" (1983, 3). They go on to argue that Third World housing agencies *undeniably* lack the finances to house all poor families at the levels adopted in the 1950s and 1960s (my emphasis). They group developing countries into three categories: those who do not have the resources for conventional standards, those who "perhaps have the resources" but have "left control over the economy in private hands or have devoted only a tiny share of public resources to housing" (1983: 2) and those who could increase public expenditure on housing but not enough to house everyone.

This widely accepted drawback of conventional housing has been the justification for the evolving forms of self-help such as upgrading and sites and services projects. Payne argues that the past experiences of individual examples "realised that a more appropriate course of action was to spread limited resources more widely - *if thinly* - and provide serviced plots in which mutual self-help could be used to construct individual dwellings" (1984: 2, my emphasis). Similarly critical of the public sector, Mehta and Mehta (1988: 127) argue that in India the inefficiency of the delivery system by the public sector is due to the problems which occur between planning and implementation. This has resulted in the patronage of middle and higher income groups rather than in the provision of housing for the targeted groups, low-income groups (LIG) and economically weaker sections (EWS).[1] In Lucknow beneficiaries of conventional housing policies are generally not of the same socio-economic standing as the squatters for whom the projects were constructed (Sinha 1991). The occupation of conventional projects by middle and higher income groups is a common feature in state-sponsored housing policies in India. Housing shortages in the middle income sector and the distant location from low-income sources of employment are two factors for this phenomenon (Bhattacharya 1990; Sinha 1991).

Siddiqui and Khan (1990) identify four **main** reasons for government-sponsored housing failing to reach the urban poor in Pakistan: targeting, affordability, policies and procedures for allotments and time-lag between allotment and the actual development of fully serviced plots. The incremental development scheme - Khuda Ki Basti[2] - initiated by the Hyderabad Development Authority has been a recent policy effort to overcome such criticisms of state-sponsored housing development through the simplification of procedures, dismissal of adherence to standards and layout plans and a prescribed involvement of middlemen in certain roles (Siddiqui and Khan 1994). The rolling back of the state's responsibility in housing provision in such a programme coincides with the wider self-help developments which seek cooperation between the formal and informal sectors in overcoming state inefficiency.

State-sponsored self-help has been criticised on a number of grounds with regard to the effects upon the political economy of the household. Burgess (1976) and Ward (1982) argue that there is a double exploitation within self-help policies which creates new markets for building, land and capital that lie in the interests of landed capital.[3] This double burden is a result of the unaccounted labour and time input of people which are required for the housing process to sustain itself. While one level of exploitation occurs during the first shift or working day during which time income is earned, the second level of exploitation takes place during the second shift in the evenings and during weekends. The result is an extension of working hours and a double exploitation of self-help builders and workers (Mathey 1992). Moser (1992) takes the analysis of exploitation through self-help further in her study of women in self-help projects. She argues that the triple role of women (productive, reproductive and community managing work) is not adequately perceived, and non-productive activities are therefore not valued in development planning strategies. While the increasingly popular 'women in development' stance has begun to occupy the centre of development debates in recent years, its ideological overtones have largely remained within the self-help framework. Despite this tendency to align itself with the self-help approach, 'women in development' advocacy has shown to be a valuable lobby group against the exclusion of women in the development process.

Ward (1982) notes in the early 1980s that few had spoken against the 'logic' of self-help, due to the lack of empirical evidence of its successes or failures at that time. A number of recent empirical studies with regard to commodification processes have since enhanced this traditionally theoretical argument given the historical experience and hindsight of self-help policies of the past decades. For example, Ramirez et al (1992) studied the consolidation of commodification in the barrios of Caracas, Venezuela where it was found that exchange relations had evolved in an irregular manner. While the access to land had remained constant and predominantly outside of market mechanisms, the ability to use land had become a monetarily governed activity. Amis' study of Nairobi's informal housing market has revealed long-run patterns within the city's low-income rental market which counter the commonly-held "'bogey-man' perception of landlords in housing markets in general and in Nairobi in particular" (1996: 281).

The critics of self-help argue that self-help is not a new found freedom as defined by Turner and Fichter's (1972) "freedom to build" but in the fact that the poor have no other options by which to house themselves (Harms 1976; Ward 1982; Burgess 1992). The most vocal opponents have been the small, but influential pool of neo-marxist researchers who have argued that the activities of the state in applying self-help models of development result in the further integration of market forces in urban poor settlements which inherently are exploitative to the poor in converting use-value to market-value. The effects of exchange or market-value conversion of housing elements are multi-faceted. Land use patterns have been drastically altered in many urban contexts whereby governments have induced commercialisation processes by encouraging the private sector to become involved in low-income housing development in what had previously been largely informal arrangements (Baross and van der Linden 1990; Angel et al 1983). Upgrading is another aspect of state-sponsored self-help policies which has led to the commodification of informal land and housing systems oftentimes resulting in the expulsion of the poorest households and in the occupation by higher income groups (Nientied and van der Linden 1982; de Witt 1992). Finally, conventional housing policies which have been most popularly criticised by the self-help school (Skinner and Rodell 1983; Turner 1972; Payne 1984) have also shown to be susceptible to commercialisation. Where complete housing structures have been built by government authorities as public or social housing, there is evidence that the accessibility of housing in such projects are not isolated cases but are determined by the wider local housing market (Mathey 1990; Hamberg 1990).

Counter to the trends in other developing countries, in India the construction of multi-storey, prefabricated systems by the appropriate state agencies has until recently been the predominant form of housing policy. The ascension of state-sponsored self-help policies has coincided with the demise of the socialist planning model in India and has reached Amritsar in the form of upgrading and core housing developments. In this respect, the case of Amritsar shows a departure or tardiness in following the wider trends in Third World housing policies with many countries such as Pakistan, Kenya and Thailand having experienced sites and services and upgrading as early as the 1970s. The adoption in India of the National Housing Policy in 1988 was a considerable shift in the direction of self-help which has since been further developed by a variety of state-sponsored self-help housing schemes.

Conventional and Non-Conventional Housing Policies

In most developing countries the impetus for state-led housing construction did not take place until after expulsion of colonial rule. After the independence of many developing countries from colonial rule, the post-colonial governments which were established faced a challenge of planning for the future development of their countries.[4] The shift away from conventional forms of housing began with the growing discontent with public housing schemes in Europe and America among policy-makers and conservative politicians during the 1960s and 1970s. This meant that the prospects for the further development of conventional housing in developing countries faced opposition. The stereotypical tower-block flats of New York City and London which had come to be associated with social degeneration were transferred to the situation of developing countries where, despite the growing gap between the demand and supply for housing and the lack of investment into social development through colonial rule, the housing problems were perceived to be exaggerated rather than remedied through conventional means.

The evolution of international housing policy trends is reflected in the development of national and state housing policies in most developing countries. As was introduced in Chapter One, the theoretical and ideological debates in the Third World housing literature have resulted in a number of types of housing practice. After three decades of extensive housing research and project implementation, the experiences of the various emerging strategies present insights into the role that housing policies play not only upon formalised systems, but also upon the informal, unregulated systems. The link between theory and practice has, as will be presented in this section, continued to exist at the forefront of housing debates and in the development of policy.

The relationship between state intervention and self-help has been subject to much debate in the low-income housing literature (Tait 1997; Ward 1982; Mathey 1992). The theoretical debates around self-help housing can be best illustrated in practice through conventional and non-conventional housing policies. Conventional policies are those which promote housing projects constructed by the building industry through state subsidies and no self-help efforts, such as those, until recently, implemented in India by the state housing boards as the most popular form of policy. Such projects tend to be planned, designed and built by both public and private agencies for social and public housing. Non-conventional policies, on the other hand, consist of a range of different types of projects which combine state assistance with self-help and involve users in the processes of housing production and exchange (Harms 1992). While the state's role and that of the individual and collective have been seen in opposition to one another, Harms (1992) has constructed a continuum of the range of possible self-help and state activities. On one end of the spectrum, he places unaided self-help and on the other end he places conventional policies (Table 2.1).

Table 2.1 Continuum of Self-help and State Housing

Unaided Self-Help	State-Supported Self-Help	State-Initiated Self-Help	Conventional Policies	Housing

Source: Adapted from Hans Harms (1992), pp. 35-36.

In this continuum 'self help' is state-initiated and 'self-help' is state-supported. State initiated self-help is reflected in sites and services projects while state-supported self-help has recently been implemented through legalisation and upgrading schemes. State-supported self-help has gained popularity due to its relatively affordable methods of dealing with the immediate needs of those who do not have access to market-supplied housing products. The state's interest in addressing the housing crisis in this way has, as some argue, been to confront it before it becomes a possible point of social unrest (Castells 1977; Alan and Gilbert 1982).

The continuum illustrates a framework which is derivative of the self-help-neo-marxist debate that was discussed in Chapter One. The intricacies of the debate have both theoretical and practical implications. Burgess, one of the most vocal neo-marxist scholars, accuses Turner's approach to self-help as paving the way for political patronage of the poor. While Turner sees self-help as a means of resistance to poverty, his analysis falls short of viewing housing as a means for mobilising the poor in resistance to economic and political structures. He prescribes the state's role as a facilitator rather than a provider of housing to the poor. Burgess, on the other hand, gives more agency to the potential power of the poor and their lack of basic needs as a mobilising factor for social action. Pradilla (1976) similarly criticises self-help housing advocacy as an individualistic approach which fails to recognise housing as an element of social reproduction bound by the capitalist modes of production. Pradilla goes on to argue that the opposition to self-help advocacy would be through the demand for housing for the labour force as part of the price for labour contributions. Despite the strong opposition by the small group of neo-marxist researchers, the self-help school has maintained the dominant position within the formulation of housing policy. As a result, the past few decades have seen the proliferation of planned interventions reflecting the middle range of Harms' continuum.

Two specific policies, sites and services and upgrading, emerged during the 1970s and 1980s, which many attribute to Turner (Marcussen 1990). Sites and services is an umbrella term for projects ranging from "plots of raw land serviced with some shared facilities to individually serviced plots with partly finished houses on them" (van der Linden 1986: 16). Upgrading is sometimes included within this term, while projects which offer no houses or parts of houses generally fall under the same category as well (van der Linden 1986). Therefore, the term covers a wide range of project types which provide varying degrees of services but none of which provide complete houses (McAuslan 1985).

The main appeal of sites and services to Third World countries and planners has been that they stimulate the housing supply by reducing the cost of housing units while also "cleaning up" localities and giving governments the right of control of the areas (Payne 1984: 2). This strategy came about after the 1976 Habitat conference in Vancouver when it was widely accepted that slum clearance and squatter relocation schemes

26

had more detrimental effects than positive ones and that anti-urbanisation and population methods had been exhausted in the context of rapidly growing Third World cities.

It was envisaged that sites and services would be the bridge between the increasing gap between housing demand and supply while also proactively tackling the underlying issues of the problem. Legal and serviced housing, it was planned, could now be more accessible to people who had been previously excluded from the legal housing market (van der Linden 1986). The idea of sites and services drew inspiration from the Abrams-Turner-Mangin school which questioned the official negative view of slum and settlements. They posed that "people know best of their own housing needs and that the government's role should be to support people's initiatives by providing basic infrastructure such as cheap land, security of tenure, and basic amenities" (Payne 1984: 2).

The sites and services which was inspired by Turner's self-help school evolved into another form when such large-scale agencies as the World Bank adopted it in place of its previous anti-housing projects (van der Linden 1986; Swan 1983). Sites and services projects became highly bureaucratised and reliant upon the World Bank to pass and implement them through its own bureaucracy as well as through those of national governments. This went against Turner's opposition of heteronomous structures in the housing process. What resulted was not a process of self-help housing schemes which aided community and individual initiatives but of top-down schemes which were dependent upon one of the largest international and economic bodies and which hardly had anything but a financial relationship with the target groups. As a result, by the mid 1980's most governments had stopped all sites and services projects because of financial problems and of ineffectiveness (van der Linden 1986).

With the widely acknowledged failure of sites and services programs to supplement the inadequate housing supply in many Third World cities, alternative approaches were sought. In order to overcome the destructive nature of clearance and resettlement and the ineffective administrative apparatus to manage sites and services, upgrading alongside legalisation were introduced as more economically viable policies both to the urban poor households and to the governments (Wegelin 1995). Rather than uprooting communities and relocating them to generally inconvenient locations, this approach meant that residents in squatter settlements could be made eligible for leases or land titles in their existing locations. The implications of this were that with security of tenure, households could proceed towards improving the physical conditions of their houses. The appropriate government authorities were also to provide basic amenities to the colonies.

Upgrading has been strongly supported by the World Bank, United Nations Development Program (UNDP) and the Asian Development Bank who, in their advocacy of the self-help approach, have welcomed this as a more cost-effective policy than sites and services. However, the details of the plans have revealed that the balance between costs and cost recovery still remain the highest priority in the planning stages (Yap 1982). As land was to be the main source of direct income from the upgrading programs, doubts have been raised regarding the ability of local governments to enforce property tax regulations and also to secure payments for rendered services such as sewerage connection and water supply (Qadeer 1983). Experiences of slum upgrading have shed light on the interconnected nature of the various sectors involved in low-cost housing and that the exclusion of certain sectors can lead to many difficulties, in particular institutional bottlenecks (Yap 1982). However, the potentials of upgrading becoming a comprehensive housing policy rely not only upon the inter-sectoral aspects of housing supply, but also upon the increasing power of private interests and the links between private interests, access to government information and resources such as land (Angel et al 1983).

The period of implementation of sites and services projects during the 1970s and 1980s was a time when India's economic development program had not been fully integrated into the global institutional framework. The World Bank has been the most active financier of sites and services housing projects, and India subsequently did not adopt a comprehensive sites and services policy as did other countries such as Pakistan and Kenya.[5] In India, however, land and legalisation policies, in addition to conventional schemes, have been more prevalent housing measures aimed at the urban poor.[6]

This section has further elaborated upon the debate between the self-help and neo-marxist schools. The various theoretical positions on access to housing have been outlined in this section in their relations to the self-help debate, but more specifically to the continuum of housing from conventional to pure self-help forms. The form of such a continuum is inherently influenced by wider international trends which have shown the emergence of an increasingly homogeneous experience of Third World countries 'down-sizing' the state's role in social housing provision towards self-help programs.

Access and Legal Status of Settlements

The emergence of the self-help housing debate in the 1960s brought about interest into illegal settlements and the significance of illegality in the housing crisis in many Third World cities. The denial of access to government or conventional housing for the poor during the 1960s is reflected in the way in which the literature of that time began to focus upon illegality. While regularisation programmes were introduced through the ideological backing of the self-help school in encouraging housing development and improvement through security of tenure, the overall approach to settlement illegality has not significantly changed over the past few decades. This section will illustrate this argument by discussing the nature and consequences of settlement illegality, land tenure and further in the intervention of the state in regularising illegal settlements.

There is a general lack of clarity in the literature on low-income settlements as to an appropriate terminology. Some argue that illegality of tenure is the distinguishing mark of squatter settlements and that shanties or slum dwellings are defined by their material content and fabric (Dickenson 1983; Abrams 1964; Stokes 1990). The introduction of material standards, construction methods, location and tenure creates complications in developing a terminology which is adequate. During the 1960s and 1970s a variety of terms were used interchangeably: spontaneous, marginal, squatter and unplanned. The failure to develop a methodological framework which encompasses the complexities of variables and contexts is most typified in the widespread usage of the term 'squatter settlement' in reference to illegal housing localities even in cases where the act of squatting is not actually present (Rodell and Skinner 1983).

The term 'squatter' is closely associated with the invasion of vacant property. However, the term 'squatter settlement' has come to be used to describe settlements which have come about through processes other than invasion. In fact, it is questionable whether or not invasions are the most common form of illegal settlement formation in Third World cities (Hardoy 1986).[7] Turner and Mangin's writings were based on fieldwork experience in Peru where most squatter settlements located around Lima were the result of invasion of land (Abrams 1966). As this was the most influential piece of research in constructing the self-help argument, the term 'squatter settlement' has come to be adopted for all forms of illegal housing, and in some cases inappropriately in describing instances where the act of squatting does not exist.[8]

Though squatter settlements show the illegal relationship between residents and the land, the nature of invasion is too broad to summarise simply as squatting. Mass organised invasions of large tracts of land can be clearly distinguished from squatting on individual plots of land (Moser 1982) while the invasion of public and private land may take on distinctly different forms (Abrams 1966). Similarly, rental arrangements within settlements formed through invasion make the distinction between squatters and renters unclear (Amis 1984). The illegality of a settlement can be most simply assumed through the absence of land titles or land tenancy documents. Such documents can be absent as a result of litigation procedures over rights to the land or because of inefficient registration systems in processing the titles (Payne 1982). In the case of sub-divisions, whether or not the appropriate government authorities were involved determined the legal status of the settlement.

The failure to possess land titles is often the result of the operation of two distinct land holding systems. This is most commonly exemplified in ex-colonial cities where customary and European land tenure systems simultaneously exist (Lea 1983; Ramachandran 1991). In these cases the acquisition of land through traditional routes does not often comply with the formal registration systems. This can result in the disregarding of customary systems for formal systems adopted from European models. While there are few specific studies on the relationship between the built environment and processes of economic, social and cultural change during colonialism (King 1990), a number of studies have investigated other aspects of such processes. Alavi (1980) notes that in India colonialism created the notion of bourgeois landed property and promoted capital accumulation through the transformation of landholding systems through colonial land revenue. Javanese cities experienced similar processes under colonial rule with the incorporation of *kampungs* and other customary land into urban areas. One obstacle presented is the creation of a rentier class of traditional land owners who seek to benefit from the continued absence of security for residents (Lea 1983).

Settlements may also be illegal on the basis of their physical layout and services. If they do not meet basic planning requirements they can be labelled as 'unauthorised,' 'unplanned' or 'pirate' because of their lack of prior authorisation for the subdivision of an area. The requirements for being granted approval for subdividing land into residential plots generally entails the approval of planning standards. The nature of legal problems that are presented to self-help housing settlements are diverse and overlapping. While the criteria for attaining regularisation in a settlement

30

may, in some cases, only specify lack of legal title, in other cases the absence of services and building code requirements may also be included (Jordan 1979). The dynamics of the various factors which are involved in settlement illegality are important when addressing the types of action that can be taken and the impacts that they have on the development of illegal settlements.

While illegality is a prominent theme in the literature on low-income housing, many have argued that emphasis on legal questions only diverts attention away from the most essential issues of improving the material conditions of residents in illegal settlements (Hardoy and Satterthwaite 1989). Abrams (1964) argued that narrowly legalistic approaches towards squatter settlements would not dramatically improve their situation. On the other hand, the issue of land within debates around illegality highlights the bourgeois nature of judicial and legal frameworks which form the categories of legal and illegal (Alonso 1980). The debates around land reveal the fact that 'illegal' land acquisition violates private property rights and the state's ability to control spatial urbanisation processes. The poor who settle on land for which they do not hold titles are merely staking their position within the urban social process, linking urban demands with political awareness (Castells 1982).

Ward (1982) notes that the early works of Turner (1963), Mangin (1967), Stokes (1962) and Abrams (1964) began the shift in the approach towards squatter settlements by confronting the traditionally-held negative connotations associated with them. They aimed to emphasise the law abiding nature and aspirations of residents of squatter settlements rather than focusing on the sole issue of illegality. Since the development of illegal settlements had occurred through extra-legal processes, the illegal qualities of such settlements were underplayed while the more conservative elements of the residents themselves were highlighted. The struggles of the poor to survive in the harsh urban context provided the basis for promoting the acceptance of illegal settlements rather than their condemnation. In Abrams' writings (1964) illegal acquisition of land by the poor was deemed as a less condemning act than other illegal types of property acquisition. Meanwhile, the squatters were portrayed as poor people searching desperately for means of survival rather than as political revolutionaries threatening the stability of the city, other than in their original acquisition of the land upon which they live (Mangin 1970; McAuslan 1985).[9]

The emergence of illegal settlements has been attributed to economic and legalistic shortcomings. From both of these angles, there is an assumption that poor people prefer to act within the boundaries of the law, given the amount of security that legal settlement provides. Therefore, the emergence of illegal settlements is a residual result of people's denial of access to legal forms of housing and land. In economic terms this implies that legal forms of housing are simply too expensive for the poor to afford to buy (Grimes 1976). The inaccessibility of mortgage finance systems further deepens the marginalisation of the poor from legal methods of housing acquisition (Abrams 1964). Similarly, speculation and tax evasion by the wealthy cause land prices to increase and remain out of the reach of the poor (Dwyer 1974). Implicit in these arguments is the belief that the appearance of illegal settlements is largely due to the failure of government authorities to control land prices through the curtailing of speculation.

Criticism of government authorities extends to their failure in implementing existing legal frameworks which govern property controls. Property laws laying out controls on urban development in this manner have also contributed to the emergence of illegal settlements. The coexistence of different legal systems in operation in the same city may also cause further discrepancy between legal controls and their implementation. As has been argued in the literature, the illegality of settlements may be the result of the inadequacy of administrative bodies to mediate customary land, sub-division control and building codes which has led to a corruption of western-oriented urban development planning codes (McAuslan 1985).

Another concern within the literature is the relationship between the state and illegal settlements. Toleration of illegal settlements by the state, it has been argued, is explained by the scale of the phenomenon which makes government authorities helpless in intervening (Dwyer 1974). Others argue that illegal settlements serve the state's interests as they are facilitators of capital accumulation and the maintenance of the status quo (Gilbert and Gugler 1982; Gilbert 1992). The state's interest in permitting illegal settlements has certainly been found to be the case in several studies where it has been a means of gaining politial support from the poor through patronage (Moser 1982; Desai 1995, Castells 1982; Cardoso 1983). A rationale given for the state's tolerance of illegal settlements is that they provide cheap housing for those who cannot afford conventional means of housing. They therefore form alternative sources of housing which absorb the demand that the formal markets are not able to provide (Gilbert and Ward 1982). In this way, illegal settlements save the government expenditure on housing and service provision. In practice, however, many illegal settlements do receive services despite their illegal status either by the state in its political patronage of the poor for support or through means of illegal connection to urban services (Gilbert and Ward 1985). Often, the phenomenon of 'stolen' services forms the basis for regularisation by the state authorities as formal service arrangements provide revenue through the charges for services provided.

For residents of illegal settlements the lack of security of tenure can cause vulnerability to abuse and manipulation by the state authorities, politicians and bureaucrats. Several linkages have been made in the literature between settlement illegality and residents' activities. The presence of "professional squatters" in several Latin American cities reveals how these illegal settlement residents can shift from one settlement to another by selling their houses and land or by receiving compensation from the government for their removal (Moser 1982). The commercialised forms of illegal settlements have shown evidence that original squatters are gaining cash reward for their bravery in illegally occupying land (Amis 1988).

The relationship between illegal settlement and housing improvement was one of the earliest arguments presented in favour of state-supported self-help. The absence of land titles, the importance of tenure security and the improvement of material conditions in self-help housing settlements was first emphasised by the founding self-help writers (Abrams 1964; Turner 1972). The dominant argument that has surfaced is that lack of security of tenure in illegal settlement deters residents from investing in the construction of permanent, solid housing construction (Turner 1972). However, van der Linden (1983: 71), in a comparative study of goths[10] and squatter settlements in Karachi, challenges such assumptions and notes how squatter households regularly invest their savings in improving their houses, despite their low incomes and their "less than favourable circumstances of tenure." A similar finding was made in Indonesia where it was noted that kampungs serve as a source of capital investment and long-term family security for households (Patton and Subanu 1988).

Settlement illegality and its significance upon housing improvement and further low-income housing development and access has been a central theme within the low-income housing literature. While illegality has been a primary concern of the policy implications of self-help initiatives, the variety of housing systems that have evolved in different Third World cities over the past few decades has resulted in a broadening analysis (Ward 1992). Self-help housing had emerged as the solution to urban development in the early 1980s at a time when sites and services, upgrading and core housing projects were being propagated. The evolving nature of land and housing markets has since created an even more complex urban context from which to establish an understanding of the dynamics of low-income housing. The exclusivity of state and market solutions to housing problems is increasingly becoming obsolete whereby the interrelated nature of market and state processes deems each to be reliant upon the other both in centrally planned and market economies (Mathey 1992). At the same time, the urban and global contexts of land, legality and settlement formation have undergone dramatic changes calling for new ways in approaching shelter issues. The next section attempts to introduce the changing nature of low-income housing markets and presents the case for an analysis outside of the limited legal framework.

Policies advocated by most Third World governments before the 1970s involved the clearance and displacement of squatter settlements. It was not until the mid-1970s that eviction was questioned and utilisation of the existing housing stock was seen as an acceptable alternative (Ward 1983). At the same time Third World governments began to accept land tenure regularisation as a viable official policy, mostly through upgrading programs (Angel 1983).[11] The government of Peru had introduced a land tenure legalisation scheme in the early 1960s through sanitation provision and legalisation of marginal barrios (Collier 1976). Under this scheme squatters were made eligible for land titles once they had paid for the installation of structural remodelling and services. The main thrust for the government to introduce such a scheme, however, was in its likelihood to reduce social unrest and to incorporate the urban poor into the ideology of the state. Private home ownership and security of tenure were used as the tools for appeasement (Lloyd 1980).

An institutional policy initiative which has evolved from Turner's argument in favour of security for housing improvement is that the documentation which goes along with legalisation is the means by which investment in settlement consolidation can be attracted (Linn 1983). However, there has been evidence in Peruvian *barriadas* that consolidation has taken place due to residents' perceptions of official intervention even where formal documentation for legalisation has not been given (Lloyd 1980). Turner viewed security of tenure as the vehicle for housing improvement, though he also saw the determination and resistance of squatters to eviction and the solid construction of dwellings as another means by which security could be achieved (Turner 1972). The varying experiences of regularisation do not always corroborate with his argument that security leads to housing improvement where the resulting situation can have negative effects upon the poorest groups, as particularly illustrated in many Latin American examples. This lack of clarity on the role of land titles in the regularisation of squatter settlements has led many to be critical of his argument and has created a particularly large gap between neo-marxist and liberal writers (Burgess 1982).[12]

Concerning the effects of land tenure regularisation, Grimes (1976) argues that it results in the improvement of service provision and encourages settlers to adapt to urban life. However, he goes on to point out that the granting of security of tenure often results in certain groups becoming beneficiaries who are not necessarily the poorest groups. Dwyer (1974) argues that regularisation through the granting of tenure can strengthen family relations while Abrams (1964) and Mangin (1970) see that it also contributes to the diminishment of urban unrest amongst squatter settlers who become home owners and thereby adopt more conservative political beliefs.

In the 1980s several studies emerged which specifically investigated the subject of land tenure (Angel et al 1983; Skinner and Rodell 1983). One point which remained unclear, even through these studies, was whether or not regularisation policies included the provision of services along with the granting of formal tenure rights. The application of regularisation policies have shown to have diverse results, with some governments failing to reveal how extensively land titles were granted to squatters (Lloyd 1980; Jordan 1979).

A theme which has been illustrated in the literature is the fact that the possibilities of regularisation policies are by no means singular (Ward 1983). In Latin America regularisation has taken on the form of the granting of ownership rights with land titles, for example in Mexico, Columbia and Peru (Strassman 1982; Moser 1982). However, studies of other countries have revealed that this is not the only method. Regularisation in African and Asian cities has commonly appeared through the granting of long-term leases, for as long as 99 years in Karachi, rather than land titles. Under this form of regularisation, residents sometimes have the choice of purchasing the title to the land, though restrictions on the transference of leases may be imposed (Baross 1983). A perhaps more widespread form of regularisation has been a reduced level of formal intervention as a means of offering security of tenure to residents. This method can be interpreted as the government's commitment not to allow redevelopment schemes to operate in the area which would affect residents. In Bombay, pressure groups have been established to lobby in favour of squatters and pavement dwellers who are threatened by eviction (Desai 1995). In Caracas legal agencies have been set up for the purpose of assisting residents with legal advice and other services, and in some cases have been successful in achieving compensation for displacement and costs for eviction (Gilbert and Ward 1985).

The regularisation of settlements has not produced definitive guidelines as to how security of tenure is perceived by residents and public/private interests. Settlements located on public land generally have fewer problems in becoming regularised than those which are on private land, provided that official agencies are committed to the schemes. Settlements on private land, however, have had different experiences with regularisation schemes. In some cases, settlements have developed as a result of short-term leases with private owners. In other cases, they have emerged through illegal squatting on private land. In both instances, there has been resistance to the transference of ownership through titles to the occupants where regularisation would threaten the current tenure systems (Baross 1983). For residents this can often result in the increase of rents or the payment of bribes to 'middlemen' in order to obtain titles. Similarly, where the financing of upgrading has been linked to the purchase of land titles, the dangers of inadequately calculating demand for legal tenure can lead to the abandonment of the program altogether. The emphasis of integrating cost-benefit analysis with regularisation can also lead to the further marginalisation of certain groups.

Overall, regularisation and the granting of tenure has offered valuable insights into the diversity of tenure structures. While illegality may cause insecurity for some households, the granting of tenure can, for others, mean an increase in housing costs. The dichotomy between illegal and legal comes into question once the prospects for regularisation are conceived. Residents are capable of detecting the comparative assurances between formal and informal guarantees of tenure security. Where informal assurances have adequately served the housing needs of poor communities, the demand for formal arrangements becomes less under demand.

Unregulated Housing Sub-Markets

Perhaps one of the most significant developments in understanding how the poor come to access accommodation has been in the extensive research done on different Third World cities. What has emerged from these studies is an acknowledgement of the diversity of sub-markets from which poor households gain access to housing. While the debate between Turner and Burgess (Ward 1982) has emerged as a hallmark of the political nature of housing studies, the diversity of urban contexts and ideological approaches within the literature, which have since been published, reveal a shift towards research on unregulated housing sub-markets (Shakur 1994; Payne 1989).

The variety of ways in which the poor find accommodation in Third World cities became a prominent topic in the late 1980s (Hardoy and Satterthwaite 1989; Kumar 1989). This was primarily due to the growing inappropriateness of the stereotypical 'slums and squatter settlements' approach which had become an oversimplified and often inaccurate description of the low-income housing systems in particular Third World urban case studies (Kumar 1989). The realisation of the complexity of low-income housing markets has resulted in empirical, micro-level case studies gaining significance within the body of knowledge around the unregulated housing sub-markets which acknowledge the variety of ways in which the poor are accessing some form of housing.

The variations between different unregulated markets are perhaps what make a singular definition difficult to construct.[13] For example, the experience of land sub-divisions in Karachi shows that illegal public land acquisition by private interests have formed a significant low-income housing market in the city (van der Linden 1983; Yap 1982). In Nairobi the growth of the rental sector in squatter settlements has been identified as a consequence of commercialisation processes in unauthorised housing (Amis 1984 and 1988). Studies on Mexico City (Ward 1976; Gilbert and Ward 1982), Calcutta (Roy 1983), Delhi (Mitra 1990) and Bangkok (Yap 1996) similarly exemplify the diversity of low-income housing experiences over the past decades. Such studies, while widening the definition of the unregulated sub-markets that are active in providing the urban poor with housing in different Third World cities, also exemplify the evolving nature of the low-income housing stock.

The unregulated housing sub-markets deliver between one and two-thirds of the total housing supply in many Third World cities, showing a steadily increasing significance on a global level in the provision of housing to the urban poor (Payne 1989). Market analysis of the unregulated housing markets has revealed a number of interesting trends. In Lagos the encouragement of home ownership among the poor has further integrated commercialisation of low-income housing markets. This has resulted in the emergence of two phenomena: "entrepreneurlandlordism" and "petty landlordism" which, while resolving the immediate housing needs of the poor, reflect a process which cannot be halted (Aina 1990: 98). Another African study of Nairobi's rental sector in squatter settlements has presented a counter-argument to the strongly upheld self-help position that the consolidation of squatter households results in more self-build activity (Amis 1984). On the contrary, the study reveals that squatter households are becoming increasingly absorbed within a growing rental sector. Amis' later study of the commercialised rental sector in Nairobi notes that unauthorised housing is becoming a marketable commodity whereby landlords are able to sell blocks of rented flats. He notes that such housing markets are "not supposed to exist, according to the conventional view...Capitalism's ability to transcend social barriers and situations and revolutionise the productive forces together with its inherent exploitation are well illustrated in the shantytowns of Nairobi" (Amis 1988: 254).

A more recent study of the so-called "booming economy" of Bangkok shows that a rapidly expanding economy and the active role of the private sector in low-income housing provision have, in fact, contributed to the displacement of the poor from their land and homes (Yap 1996: 322). The new housing stock being provided by the private sector, despite its availability, is not suitable to the locational needs of the poor which has pushed them to the peripheries of the city. Due to speculation, the low-cost housing units have become occupied by middle and higher income households.

The historical development of unregulated housing markets in specific urban contexts is another basis for the diversity of processes and trends. Several studies have relied upon the historical evidence of land use patterns, spatial segregation, legislation and political action in discussing the present position of the unregulated housing markets. Low-income housing in Karachi has been most extensively investigated by a number of Dutch researchers since the early 1980s (van der Linden 1983; Schoorl et al 1983; Nientied et al 1982). While primarily focusing upon the causes and characteristics of illegal settlement development, such micro-level studies have also brought under scrutiny the dynamics and historical evolution of squatting. One of the most distinguishing features of illegal settlements in Karachi is in their location upon public land. The influx of partition refugees in 1947 and again from Bangladesh in 1972 contributed to the heightened pressure upon land and infrastructure in housing the poor. Due to the demand for and scarcity of public land, the processes of consolidation and development of such settlements through illegal subdivisions has shown an increasingly commercialised market for low-income housing which is making squatting increasingly difficult.

Cross-cultural studies on Asian, African and Latin American contexts have offered scope for comparative analysis of the unregulated housing sub-markets. Research on Latin American cities has provided perhaps some of the most significant contributions to the study of unregulated housing sub-markets. Ward (1983) traces the growth of irregular settlements in Mexico City to the 1920s after agrarian reform in 1917 created a system of redistribution of land rights to *ejidos*.[14] The demand for such land increased over time and many *ejidatorios*[15] found themselves as owners of land which had little agricultural value but was in commercial demand due to the expansion of urban areas (Ward 1983: 37). Subsequently, the *ejidos* have formed the basis for illegal subdivision and the growth and concentration of irregular settlements in Mexico City.[16]

A more recent study of Mexico's urban housing environments investigates the historical transition from import substitution to export-led growth and industrialisation (Pezzoli 1995). Due to the government's emphasis upon higher production levels and greater competitiveness of the labour market, wages have declined resulting in worsened living standards for the poor. Gilbert (1992) similarly argues that the economic recession has had negative impacts upon housing and land tenure, services, plot sizes and eviction threats. A rising trend has been noted in both Pezzoli and Gilbert's studies that macro-economic growth strategies have shown not to encourage further self-help construction but rather to increase the proportion of renters to homeowners (Pezzoli 1995; Gilbert 1992).

In Calcutta the partition of Bengal in 1947 resulted in an influx of refugees from East Pakistan (now Bangladesh), creating an immediate demand for housing. The sudden rise in land values encouraged landlords to evict *thika* tenants and to sell their land for considerable profits. The state responded with a series of tenancy acts which sought to protect the rights of *thika* tenants. However, the protection of *thika* tenancy had a counter-effect of halting the development of land for more *bustees* by landlords, thus limiting the low-income housing (Roy 1983).

Mitra's (1990: 219) study of Delhi's land supply markets argues that the state's role as a landlord "has demonstrated amply that public landlordism and the socialisation of access to land are not synonymous." The research revealed that the commercialised sector has played a significant role in supplying land to unauthorised colonies, accommodating approximately 20 percent of Delhi's total shelter units in 1985. In Delhi the public sector has also been an active agent in the unregulated housing sub-markets. Land ceiling and regulation were intended to control the monopoly of private land owners and developers in Delhi's land markets and also to increase government revenue through public monopoly (Mitra 1990; Raj 1990). Policies executed by the Delhi Development Authority which have been aimed at large-scale acquirement and delivery of land have not been effective in acquiring all targeted plots of land. In fact, it has been noted that they have had the reverse effect of encouraging more informal settlement and the growth of unauthorised colonies rather than in controlling them (Gupta 1985; Sarin 1983; Mitra 1990).

While there is a sparse representation of sociological studies on unregulated housing in Third World cities, a number of recent studies nonetheless have shown an interest in social research on unregulated housing markets. A study of poverty and social networks in the *katchi abadi's* of Pakistan deployed the increasingly popular techniques of Participatory Rapid Appraisal (PRA) to disaggregate definitions of poverty (Beall 1995). Qualitative methods were used in ten micro-level case studies to assess residents' perceptions of their vulnerability, structural position and reliance upon social networks for security. A study of participant observation and qualitative research evaluation of housing in Lucknow sums up the inadequacy of urban low-income housing research in producing sociological and social investigative research. Sinha argues that

> the impact of housing on social and cultural parameters of society is generally lacking. While it is recognised that there are no universal solutions to the housing problems in the developing world, systematic studies of how housing schemes perform in widely differing economic and cultural contexts are needed (Sinha 1991: 33).

The unregulated housing sub-markets show a diversity of housing experiences of the urban poor in various Third World cities. Analysis of the unregulated housing sub-markets is a relatively recent trend within the low-income housing literature, and the direction of further developments of such studies remains relatively uncertain. The proliferation of these sub-markets is a direct outcome of the prolonged dominance of self-help housing policy and commercialisation processes showing the increasing importance of extra-legal and non-formal means of housing supply. The next chapter will pause momentarily from the housing debates and will turn towards the city of Amritsar which is the site of the empirical dimension of this study. In presenting the local socio-economic history of the city, parallels will be drawn between the city's history of decline and static development to that of the position of the poor.

[1] The use of such categories for targeting groups for social housing will be further explained in India's national housing policy in Chapter Five.

[2] Literal meaning: "God's colony."

[3] See Kosta Mathey (1992), *Beyond Self-Help Housing*, Mansell: London, 383.

[4] For example, colonial settler government's policies in Africa towards local urban labour were based on the assumption that provisions should be made temporarily and as basic as possible, also affecting the development of postcolonial housing policies (Rakodi 1995).

[5] The Dandora Project in Nairobi (Chana 1984), the Metrovilles of Karachi (van der Linden 1986) and the Incremental Development Scheme in Hyderabad (Siddiqui and Khan 1990 and 1994) were pioneering sites and services projects which, while having their own successes and drawbacks, were exemplary cases of the sites and services method of housing development.

[6] During the 1970s and 1980s the Indira Gandhi-led Congress government rejected the infiltration of foreign development agencies. For example, a large American development organisation called Care was banned from India in the early 1970s in response to the U.S. government's bilateral foreign policy with India in imposing its own investment interests as well as its neo-liberal ideological agenda. Loans which were being increasingly taken by other Third World governments from the World Bank and IMF were resisted by India along similar lines as an apparent continuation of Nehru's socialist model which resulted in strengthened ties with the Soviet Union.

[7] It is most likely that the term has persisted due to the fact that the earliest writers argued that squatter settlements were not areas of urban chaos and degeneration as had been previously been believed.

[8] Even Turner and Mangin specified the diversity of forms which arise through means other than organised invasions (Mangin 1967) such as clandestine sub-divisions where residents have bought plots (Turner 1969). There is also evidence of people purchasing land in squatter areas, particularly in Latin America (Gilbert and Ward 1985:100). The overall picture presented by writings of the 1960's and early 1970's shows the predominance of illegality as a key variable.

[9] The theme of law and order in these discussions is significant, and the tones within the literature reflect the aim by the authors of convincing the critics of illegal settlements of their benign, non-revolutionary qualities.

[10] Van der Linden (1983) refers to passively urbanising villages around Karachi as *goths*.

[11] This change in approach towards regularisation has been attributed to the work of Turner and Mangin in their study of Peru. Their research findings made Peruvian government policy an example for other Latin American cases (Lloyd 1979; Gilbert and Ward 1985).

[12] Regardless of Turner's intentions, the inception of regularisation has been closely associated with his early work in Latin America (Dwyer 1975; Marcussen 1990). Since then, liberal writers have been encouraging of the concept of land tenure legalisation, though not totally uncritically, seeing security of tenure as the means for people to improve their housing condition (Payne 1984; Nientied et al 1982). Meanwhile, neo-marxist writers are critical of regularisation and upgrading as it further incorporates the poor into the highly competitive formal housing and land markets and thereby invites commercial interests into these settlements (Burgess 1978,1982).

[13] Shakur (1994: 1) defines unregulated housing sub-markets as "all forms of settlements which do not possess full legal status." The definition goes on to specify, however, that the traditional slums and squatter settlements are not included as these are informal forms of settlement, although some forms of unregulated housing markets have in fact emerged as extensions of squatting.

[14] Agricultural communities.

[15] Beneficiaries of the redistribution of agricultural land.

[16] Gilbert and Ward (1982: 69) similarly trace the historical roots of illegal settlement patterns in Mexico City and Valencia to the *ejidal* land which had low agricultural value due to its undesirable physical conditions but upon invasion by low-income communities became subject to market prices.

3 Amritsar: the Region and the City

Introduction

The city of Amritsar occupies a unique position within the histories of Punjab and India. As a city at the corridor of historical throughways to India and a city now on the contemporary border between India and Pakistan, Amritsar's own demography is telling of the region's history. The initial inception, growth and development of the city took place as a move to consolidate the Sikh faith and community and then later to fortify the kingdom of Maharaja Ranjit Singh. Subsequently, the decline of the city was a result of British colonial rule and the partition of the region of Punjab in which Amritsar was reduced to a provincial border town from the economic hub of trade and travel that it once had been. All of these changes had significance upon the political climate in Amritsar where the ethnic and religious composition of the city was dramatically changed alongside the trade and economic role that it played regionally. On a perhaps more ironic level, Amritsar was reconstituted as a city of importance on both a regional and international level due to the political repercussions surrounding the storming of the Golden Temple in 1984. This resulted in a prolonged period of state repression and anti-state violence whereby the city was immersed into yet another period of decline and stagnation. The chapter presents a picture of the structurally marginal position of the poor within a city which has itself been structurally marginalised to development processes in the region.

Historical Background of Amritsar (1577 to 1849)

The city of Amritsar is one of the central towns in the corridor to India for those travelling by land from the north-west. Located on the Grand Trunk Road which runs from Calcutta in the east to Peshawar in the west, the city has seen numerous invaders over the centuries who have passed through the town on their way to Delhi. The urban history of Amritsar has been

43

strongly influenced by the economic and political development of the region. Amritsar is a cultural and religious centre not only in the state of Punjab but also in South Asia. Founded over 400 years ago by one of the Sikh Gurus, Ram Das, Amritsar has become the political and cultural centre of the region and the religious centre for the Sikh faith. Its main significance was that of a link city between Lahore and Delhi until the latter half of the sixteenth century in 1577 when the fourth guru of the Sikhs, Guru Ram Das, selected the site as a place to construct the *sarovar*, a body of holy water. The location for the construction of the Golden Temple had been chosen by Guru Ram Das in an area that belonged to the village of Tung. It is unclear whether the land was bought from the villagers by the Guru, whether it was given by the Mughal Emperor Akbar rent-free, or donated by villagers (Mavi and Tiwana 1993; F. Singh 1990). Guru Arjan Dev, the fifth Sikh Guru, completed the task of building the *sarovar* and also built a place of worship called the Golden Temple - *Harmandir* (the House of God) - next to it, further sanctifying the body of water.

For the first two hundred years of its history the present city of Amritsar was not known as Amritsar but as Ramdaspur or Chak Guru (Grewal and Banga 1979). During this time, Ramdaspur was a small town and in fact acted merely as a township to its neighbouring Lahore. By the end of the seventeenth century, Amritsar had become a bustling centre of trade, particularly in textiles and edible goods, and served as a satellite town to the nearby metropolis of Lahore. It was at this point that the political stability of the region of Punjab was being threatened by both internal divisions and external threats of Mughal rule. Ramdaspur became a rallying centre for political organisation of the Sikhs as well as territorially significant as a target for Afghan invasion and of independence from Mughal domination. This period was marked by political instability in the region and the town of Amritsar frequently suffered from bouts of invasion and the imposition of new rulers. Resistance to these incursions was successful in Ramdaspur, and as a result several of the Khalsa, those who fought in defense of Ramdaspur, established their residences around the core of Ramdaspur and the Golden Temple. Different groups built homes and put up their own small forts surrounded by defensive walls, remnants of which still remain today and which have distinctly shaped the physical development of the city since. These communities, known as *katras*, were named after their founders and contained markets and residential areas which were centrally administered by their respective founders. Eventually, these communities became concentrated through settlement of more families from other areas. It was through these *katras* that the city of Amritsar came into its own as a centre of commerce and trade.

By the time the British began to make significant inroads into the Indian sub-continent, the Mughal empire was in decline and Punjab was in the hands of a large number of Sikh *misls* (warrior groups) who were often in dispute with each other. The wranglings between various Sikh groupings were resolved by the nineteenth century when Maharaja Ranjit Singh unified the Punjab from Peshawar to the Sutluj River. Amritsar came to be known as the second capital of his empire, second only to Lahore. It was through Ranjit Singh that the *katras* became unified into what is now Amritsar.

During his era of rule in Punjab, the city of Amritsar expanded and intensified. Economic growth within the kingdom of Maharaja Ranjit Singh as with other princely states in other parts of northern India became concentrated upon defense measures, the building of architectural structures and conspicuous expenses of the court. Threats of attack encouraged Ranjit Singh to enclose the city with a wall to make it more easily. At the time of the annexation of Punjab by the British in 1849 Amritsar was the largest city in Punjab. Due to colonial policies, however, Lahore soon became the largest town in the region. This marked the beginning of a period of decline in growth and political and economic significance (Gauba 1979).

Colonial Rule and the Partition (1849-1947)

The British impact on urban areas in South Asia was significant. Punjab did not experience the levels of urbanisation and industrialisation during this period that the colonial metropolitan port cities of Bombay, Calcutta, and Madras did, resulting in the decay of already established towns and cities outside of the main urban spheres. However, the military significance of Punjab - in terms of recruits to the British Indian army - meant that army housing and facilities dominated many urban centres in the region. Cantonments and civil lines were concentrated in Punjab and Uttar Pradesh to house British officers and their families and men of the armed forces. Indian soldiers were housed separately within the cantonment areas. During British rule, Punjab served as a predominantly administrative, though towards the end of colonial domination had become a considerable industrial and commercial, centre. Amritsar, in particular, had begun transforming into an industrial city between 1911-1921, primarily through the textile industry. This expansion continued until 1931, at which time there were approximately 600 looms employing 3000 weavers in the town (Sandhu 1989:26). The expansion of the textile industry and trade in Amritsar created a rapid demand for labour which resulted in the growth of the city from in-migration. However, the economic depression in the West halted the demand for textile goods which had previously been exported to Europe and the United States. As a result, the first phase of the city's decline began.

The northern side of Amritsar was developed as the residential area for British officers and their families. This part of the city is now occupied by Amritsar's upper and upper-middle class families where there are wide roads, fountains, gardens, and shopping areas. This is in sharp contrast to the congestion of the walled city. This spatial segregation, created by the British, was strictly enforced and was to have an impact on the spatial and social development of most colonial cities even after independence (Ramachandran 1991). Hazlehurst (1968) has noted that the area north of Delhi and westward to Punjab is generally characterised by middle-sized cities which are found along road and railroad routes that had been developed by the colonial administration, a common remnant of colonial urbanisation. The concentration of trade and commercial activities within these cities is largely along these transport routes (Rondinelli 1983). Amritsar's location upon the Grand Trunk Road made it a prime location for colonial administrative units, providing a route to other parts of Punjab and the sub-continent.

Although the influence of British imperialism manifested itself differently in Punjab's urban and rural development in comparison to other regions of the sub-continent, it is of utmost importance to any study of social development in the region to address not only the long-term inequalities that were created by the empire's economic exploits through its port cities, but also its segregational policies through its administrative centres[1] (King 1976). Punjab is a prime example of the latter type of region which had many pre-existing physical divisions such as the walled city and the katras which were further reinforced by colonial administrative practices. The cantonment and civil lines today mark the separation of the elite from the non-elite as they did during colonial rule between British and indigenous communities. The segregation of communities along economic, class and caste lines has continued to dominate the spatial development of Amritsar.

As a result of the partition of Punjab in 1947, Amritsar was one of the worst affected areas as it had, previous to partition, been located in the centre of the province and thereafter found itself on the border of a divided Punjab. Having once been at the crossroads of central Punjab and a major economic, religious and cultural centre, Amritsar became the last city on the Grand Trunk Road before the border demarcating India and Pakistan.[2] The economic life of Amritsar suffered greatly as a result. While Amritsar also became the terminus of the East Punjab Railways, trade between the two sides of Punjab became strained because of trade restrictions and enforcement of income-tax clearance requirements. The area to the west of Amritsar became, in practical terms, very distant, and a move towards developing economic links with the rest of East Punjab and India were inadvertently imposed. Raw materials which had previously been obtained from towns in West Punjab and chemicals and machine goods from Karachi were now only available from Bombay. The textile industry among other industries in Amritsar which had historically been significantly reliant upon the northwestern region was now forced to search for markets in the Indian interior (Gauba 1988). Thus, the importing of raw materials, machinery, and chemicals, the two-way burden of railway transport for freight, and the shifting of markets for manufactured goods became obstacles for the industrial and commercial sectors in Amritsar (K. Singh 1991).

The border imposed other restrictions upon Amritsar. Tension between India and Pakistan caused a sense of insecurity in the areas along the border which, apart from defense measures, deterred both private and public investment. Already existing businesses in Amritsar were also affected by this. A large number of industrialists and traders expressed a desire to move to other cities and even other states (Sandhu 1989). In an effort to prevent the liquidation of the area's industrial base, the government passed the East Punjab Factories (Control and Dismantling) Act 1948 which put heavy restrictions on the rights of any person to move machinery or any other parts from a factory without official permission (K. Singh 1991). However, many businesses were successful in shifting their interests to other cities (Rai 1986). Similarly, those refugees from the west with capital to begin new businesses predominantly opted to settle in the larger commercial centres of Delhi, Bombay, and Ludhiana rather than in the uncertain climate of Amritsar (Luthra 1949).

The demise of Amritsar as a commercial and manufacturing centre effected the poor with regard to income-generating options. Even though Punjab had been an industrially underdeveloped state with a high agricultural and small-scale and handloom industries output, the labour-absorbing economic activities which had historically been based in Amritsar were now moving to other, less insecure locations. Amritsar's commercial and industrial sectors could no longer provide skilled and unskilled employment to the poor in the same capacity. Even while smaller textile, trade and manufacturing houses emerged as the new form of productive activity in the city, the degree of job opportunities available for the poor became minimal. Post-partition Amritsar slowly became static in its economic and population growth. Economic in-migration dropped dramatically, and one result was a steadily decreasing urban population growth which eventually saw the rise of Ludhiana as the new economic centre of east Punjab. As is clearly evident in the past two decades through the growth of the city's population (See Table 3.1), Ludhiana has become the magnetic force for economic investment, industry and labour (Oberai and Singh 1983).

Since the partition of Punjab the average percentage growth of Amritsar has steadily decreased whereby it has consistently fallen below the state average (See Table 3.1). As is evident in the 1951 data on Amritsar, the city's population figures were greatly altered as a result of partition where rates of growth of urban population responded to the political turbulence surrounding the partition of the region (H. Singh 1985).

Table 3.1 Urban Population Growth of Amritsar, Ludhiana and East Punjab

Year	Total Population of Amritsar with percentage increase		Total Population of Ludhiana with percentage increase		Increase of East Punjab's Urban Population
1941	391,010		111,639		N/A
1951	325,747	-16.69%	153,795	37.76%	20.02%
1961	376,295	15.52%	244,072	58.70%	29.06%
1971	434,951	15.59%	401,176	64.37%	25.27%
1981	589,299	30.79%	606,250	51.12%	44.51%
1991	709,456	19.27%	1,012,062	66.94%	29.11%

Sources: Government of Punjab (1982) p. 56 in Oberai and H.K. Singh, *Census of India,* 1991.

The shifting of businesses and administrative offices after partition was the beginning of the decline of the city. In post-1947 East Punjab, Amritsar became increasingly associated with its historical significance in the region rather than in its potential as a future major city in the state. Trade in textiles and dry fruits and small-scale, hand-loom manufacturing still continued to be large economic activities in Amritsar. However, the scale of these activities in an increasingly industrialising era meant that Amritsar could not compete with cities like Ludhiana which, by the 1970s, became the largest city in both size and in economic growth.

Along with the gradual economic decline of Amritsar, the number of employment positions available became limited. This significantly affected skilled occupations where the professions of weaving, pottery making, and dyeing were undermined due to both the economic depression and as well as to the mechanisation of many traditionally manual activities. As a result of these changes, the income-generating activities that these people became involved in became diversified and forced many to take on unskilled jobs. The changes to the occupational structure created by the declining economic climate of Amritsar, in many cases, challenged traditional caste hierarchies which had previously withheld occupational mobility amongst the lower castes (Saberwal 1990). However, this did not result in a massive upward mobility of low-caste groups. Instead, there was an increased amount of occupational mobility amongst the low caste groups which also meant that the jobs of lowest pay and status became more readily available to groups which may have previously attached social stigmas to them.

The forced migration of people on both sides of the newly formed border meant that entire communities were uprooted from their native surroundings, livelihoods and homes. For many rural refugees, migration across the border involved a journey to a destination where they would be allotted agricultural land. However, for others the nearest urban centres to the border presented options of lesser distances to travel and the assurance of the economic dynamism and absorptive nature of cities. Amritsar at the time of partition was the first city across the border on the Grand Trunk Road and also the largest city in East Punjab which for many offered more hopes of economic survival than did other urban centres. The processes earlier mentioned of the demise of Amritsar's commercial and industrial bases would prove to be a disappointment for many partition refugees who chose to settle in Amritsar (Purewal 1997).

The mass migration on both sides caused the pool of skilled and unskilled labour in East Punjab to change significantly. A large number of urban Muslims in East Punjab who left for Pakistan were artisans, craftsmen, blacksmiths, and potters while a majority of the Sikhs and Hindus coming from the western side belonged to the trading classes (Rai 1986). According to government estimates the number of Muslim artisans who left for Pakistan numbered 1,854,188 and the number of non-Muslim artisans who settled in East Punjab totalled 252,873 (Government of

Punjab 1947-50). This created a gap in the labour structure which, for Amritsar, had a particularly significant impact on the textile and embroidery trades (Table 3.2).

Table 3.2 Transfer of Occupations at Partition according to Religious Groups

Hereditary Occupation	Hindus and Sikhs in West Punjab (100,000)	Muslims in East Punjab (100,000)
Agriculture	8.2	29.60
Traders	14.01	2.79
Weavers	0.08	3.70
Shoemakers	1.25	1.64
Carpenters	0.56	0.79
Blacksmiths	0.57	--
Potters	0.45	1.64
Dyers	0.04	0.41
Bakers and water carriers	0.57	1.84
Barbers	0.17	0.86
Sweepers	2.10	0.07
Washermen	0.05	0.52
Tailors	0.02	0.08
Total	28.07	43.94

Source: Kirpal Singh (1989:183).

An imbalance was created as a result of this migration of labour. The poor, low-caste refugees who came to settle in Amritsar came from a variety of occupational backgrounds (Sharma 1996). While many poor employment seekers changed professions in response to the skewed demand created in certain sectors, many were forced to take on what may have been considered comparatively menial tasks and more importantly less reliable sources of income. This resulted in a large pool of both skilled and unskilled labour competing for a limited number of jobs with much of the skilled labour force being forced to take on unskilled jobs.

The resettlement processes that would immediately follow the arrival of partition refugees, though treated at the time as a transitional stage, formed new patterns of residential localities which would set the precedent for the future. Immediate relief for refugees was prioritised over the long-term urban allocation and rehabilitation schemes that were taking place in the rural areas. The problem of overcrowding in relief camps where most refugees were being sheltered was an additional pressure for swift rehabilitation policy (Kudaisya 1995). In the meantime the unauthorised occupation of Muslim evacuee houses by local residents and, in many cases, by civil servants meant that the supply of houses for refugees became rapidly diminished before a scheme could even be developed and executed. Evicting these unauthorised occupiers became a task too large to contemplate for the concerned government authorities. By the time the survey of Muslim immovable property was finally finished in June 1948, most of the houses had already been occupied.

A further pressure on the supply of vacant houses came from the influx of landless refugees who could not secure employment in the rural areas and who opted instead to move to towns and cities (Rai 1986). For the most part, refugees were forced to live under the most basic conditions in public buildings or in make-shift structures on government land until government assistance arrived. The disturbances at the time of partition had resulted in substantial damage to houses and buildings (Kudaisya 1995). People who occupied evacuee houses were not provided with necessary repairs until years later while many of those who were only given plots of land to built make-shift temporary homes found these plots to be their permanent homes.

The overall outcome of urban resettlement policies in Amritsar was that refugees did not benefit from the partially implemented rehabilitation programs. Relief measures of temporary plot provision and long-awaited repairs on allotted evacuee houses meant that people were left to live in sub-standard conditions with very little assistance, if any. Another important point is that wealthy and influential people were able to reap the benefits of the free-for-all situation created by partition. As in the walled city, these privileged people, both refugee and local, managed to gain accommodation and land in far excess of what had been documented as appropriate allotment standards. Thus, the immediate resettlement process at the time of partition in Amritsar further deepened class and caste inequalities and divisions.

One of the most significant long-term effects of partition migration upon low-income housing in the city has been its establishment of squatting on public land as a means to access for housing for the poor. The Rehabilitation Department was incapable of mobilising resources to adequately compensate urban refugees. As a result, squatting upon public land was permitted by the authorities as an immediate solution to a large problem. Subsequently, the inability of public housing agencies to provide enough public schemes to accommodate residents in squatter settlements has similarly relied upon the availability of public land for squatting. This immediate solution to what has been perceived as a housing deficit has become a predominant path to housing for low-income households in Amritsar.

The Region of Punjab

The name Punjab is of Persian origin and means five (*punj*) rivers (*ab*). The literal definition of the region refers to its geographical location between the five rivers of the Sutluj, Beas, Ravi, Chenab, and Jhelum (Mavi and Tiwana 1993). Although Punjab now exists as two separate entities, East and West Punjab divided by the boundaries of India and Pakistan, the region has been geographically and politically divided at numerous points in history. For instance, at the beginning of the nineteenth century its boundaries stretched from Afghanistan in the west to Kashmir in the north down to Delhi. Present day East Punjab only extends from Amritsar to Chandigarh (Figure 3.1).

Figure 3.1 Punjab: Past and Present

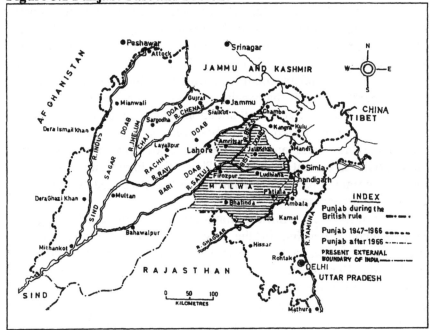

Source: Mavi and Tiwana (1993: 159).

Ethnically, linguistically and culturally there are many similarities between the eastern and western regions of Punjab. Amritsar and Lahore

52

are only forty miles apart and up to the 1947 partition were, more or less, 'twin cities.' During partition many Muslims fled into what is now Pakistan and many Sikhs and Hindus did the same to India. This significantly affected the social make-up of Punjab, creating a Hindu/Sikh majority in Indian Punjab with less than 3 percent Muslim and other religions (Table 3.3).

Table 3.3 The Changing Punjab

	Area sq. km.	Population (millions)	Muslim (%)	Hindu (%)	Sikh (%)	Other (%)
1941	256,600	28.4	53	31	15	1
1951*	122,500	16.1	2	62	35	1
1961	122,500	20.3	2	64	33	1
1966 +	50,260	11.2	--	45	53	2
1971	50,260	13.5	--	38	60	2
1981	50,260	16.7	--	36	62	2

* After the creation of Pakistan
+ After the separation of Haryana
Source: Jeffrey (1986:42).

In post-independence India, states were demarcated nominally along linguistic lines. The States Reorganisation Committee recommended that Punjab, in addition to the borders of other states in the Indian union, be reconstituted on a linguistic basis. Therefore, in 1966 the state of Punjab was further divided on the basis of language resulting in the carving out of the new states of Haryana to the south and Himachal Pradesh to the north. Sikh and Hindu religious identities were also at the forefront of the new linguistic state demarcations which, as will be discussed in the section on ethnicity and communalism below, had serious implications upon communal politics in Punjab and Amritsar (Vanaik 1990). The linguistic reorganisation of states caused the population of Punjab to be divided by half while the geographical area was decreased by two and half times (See Figure 3.2).

Contemporary Punjab has an area of 50,362 square kilometres and a total population of 20,281,969.[3] The urban population is about 14 times more concentrated than the rural areas with population density in villages being 292 persons per square kilometre as compared to 4159 persons per square kilometre in cities (Gupta, 1993: 65). According to the 1981 census, Punjab had seven class I cities:[4] Ludhiana, Amritsar, Jullundur, Patiala, Bhatinda, Pathankot, and Batala. The 1991 census reveals a substantial growth in urban population in the state with twelve class I cities (Table 3.4).

Figure 3.2 East Punjab and its Districts

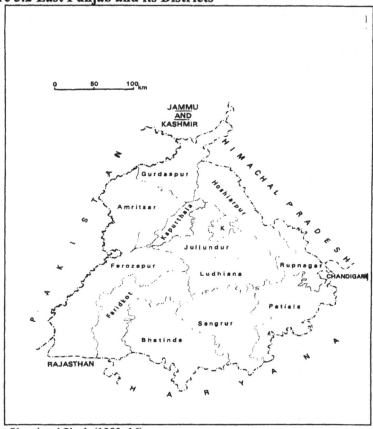

Source: Oberai and Singh (1983: 16).

Table 3.4 Punjab's Class I Cities

Urban area	1981	1991
Ludhiana	606,250	1,242,781
Amritsar	589,229	853,831
Jalandhar	405,709	728,802
Patiala	205,849	566,973
Faridkot	-	439,839
Sangrur	-	417,994
Gurdaspur	-	386,412
Firozpur	-	384,400
Bathinda	-	351,133
Roopnagar	-	232,317
Hoshiarpur	-	222,138
Kapurthala	-	166,605

Sources: Government of Punjab (1982:56) and *Census of India* (1991).

Amritsar is a medium-sized city which has experienced low levels of urbanisation. Punjab is one of the smallest states in India and has a higher percentage of urban to total population than the all-India average. Amritsar, particularly when compared to Ludhiana which is far above the national and state average, has a considerably lower level of urban growth. Table 3.5 shows the patterns of urbanisation for India, Punjab, Ludhiana and Amritsar.

Table 3.5 Patterns of Urbanisation in India, Punjab, Ludhiana and Amritsar

Census Year	1971		1981		1991	
	Urban Pop. % of total population	Decadal Growth 1961-1971	Urban pop. % of total pop.	Growth 1971-1981	Urban pop. % of total pop.	Growth 1981-1991
India	19.91	38.23	23.34	46.14	25.72	36.19
Punjab	23.73	25.27	27.68	44.51	29.55	28.95
Amritsar	29.15	15.59	31.83	30.79	33.0	19.27
Ludhiana	34.81	64.37	42.10	51.12	42.0	66.94

Source: *Urban Statistics Handbook*, National Institute of Urban Affairs (1993), Oberai and Singh (1983) *and Census of India* (1991).

There is a low level of urbanisation present in Amritsar in comparison to national and state patterns of decadal growth and total to urban population percentage. The contemporary statistics on Amritsar reveal that it is not a city exhibiting rapid industrialisation or expansion. However, its urban history reveals that its low level of growth is a recent phenomenon due to a number of factors which have affected its economic and political significance in the region. There is a stark gap in the literature in reference to urbanisation and low-income housing in Punjab. This is partly due to the emphasis in the literature on large cities, Punjab only having one city over the one million mark, but also due to its socio-political history. The emphasis on large cities has meant that areas such as Punjab and Bihar, which do not necessarily contain booming metropolises, but which have predominantly agrarian bases, have been virtually ignored by studies on urbanisation and housing.

Migration and Development

During the 1950s and 1960s agriculture became a main source of economic growth in the north west region of Punjab, through mechanisation and intensification methods.[5] The effects of mechanical and chemical inputs upon the agrarian structure were profound in displacing many poor households which had previously earned their living from manual agricultural activities. The flows of migration in Punjab have exhibited a complexity of elements and processes. Rural to urban migration has been accompanied by areas of highly concentrated in-migration from other less economically developed states such as Bihar and Uttar Pradesh. The emerging demographic and economic trends from the past few decades of rural-urban migration have shown that imbalances have been created rather than eradicated as had been commonly perceived by modernisation theorists (Lewis 1954). Rural to urban migration, in-migration from backward Indian states and out-migration, predominantly from rural areas, to places such as the U.K. and North America are three notable trends of migration in Punjab. Differential inter-sectoral income levels, Green Revolution agricultural policies and steady industrial growth have all had effects upon the flows of migration that occur. Differences in income levels between sectors has also had a significant impact upon rural to urban migration with an increasing proportion of workers engaged in agricultural and non-agricultural activities.

Bonded migrant labour, remittances to villages, marginalisation of the poorest groups and circulatory migration due to seasonal labour are a few points of analysis of the relationship between migration and economic development which have been pursued in a number of empirical studies of the Punjab (Singh 1997; Byres 1983; Oberai and Singh 1983; Gill 1991). Out-migration from Punjab has also been a significant feature which, particularly due to the technological advancement of agricultural production and its impact upon labour movement and structures, has contributed to the movement of people from rural areas in Punjab and in the development of a Punjabi diaspora (Singh and Thandi 1999, Tatla 1998).

The relationship between rural and urban economic development in Punjab reflects the planned development strategies of the Five-Year plans which aimed to transform the economy through a link between the urban and the rural spheres.[6] Urban areas in Punjab have over the past three decades experienced the modernisation model of planned transfer of agricultural economic surplus to investment into the manufacturing sector. However, the rate at which agricultural growth had been generated in the initial stages during the 1960s and 1970s was not, as some argue, sufficiently balanced with industrial growth in the region (Gill 1996).

The growth of agricultural activities through large landowners and capital-intensive methods has meant that many agricultural workers have been displaced while the demand for cheap labour has attracted workers from poor regions of India, in particular from the state of Bihar. A comparative analysis of the socio-economic indices between Punjab and Bihar reveals the imbalances between the two states which has contributed to the flow of economic migration from Bihar to Punjab. Agricultural statistics show that in 1986-87 Punjab had a per capita gross cropped area (in hectares) of 0.60 compared to 0.17 for Bihar (Singh 1997: 297). Perhaps more significant is the discrepancy between the statutory minimum wages of agricultural labourers in both states with 10 rupees per day in Bihar and 33.3 rupees per day in Punjab. The profile of poverty levels in each state (1983-84) shows a similarly dramatic difference between Bihar and Punjab with the percentage of people living below the poverty line[7] in Bihar being 51.4 percent in rural areas and 37.0 percent in urban areas while in Punjab being 10.9 percent in rural areas and 21.0 percent in urban areas. Thus, the movement of people from Bihar to Punjab in search of better economic opportunities can be seen as a response to the differing economic situations in both of these states.

The pattern of urban growth and development in Punjab has, as a result, taken the form of areas of highly concentrated industrial activity where rapid expansion of the manufacturing sector has generated a demand for labour. This has subsequently contributed to the increase of settlements, particularly low-income residential areas, in close proximity to industrial areas. The development of squatter settlements upon fringe areas as well as upon public land close to places of income generation has steadily increased whereby speculation of land values has been dramatically affected by industrial and commercial growth.

Alongside the growth of industrial production in certain cities, informal activities have also attracted many migrants from regions where unemployment and poverty are at high levels. The uneven development within the Indian union has made Punjab the 'promise land' for many poor migrants. Industrial development within Punjab has also been concentrated in particular cities such as Ludhiana and Batala. Therefore, regions outside of the areas of concentrated capital activity, such as Amritsar, have not experienced the same levels of in-migration. The competition for access to employment within the highly concentrated industrial areas has had a knock-on effect of dispersing labour to fringe areas where opportunities in informal and tertiary sectors are more accessible. In Amritsar it is this type of migration among economic migrants which is predominant.

While it is not possible to generalise about urban growth and development in India it is reasonable to develop particular themes about the various regions. In this case Punjab reflects the impact of colonialism, industrial development and in-migration from other states of the Indian Union. Each of these influences on urban development have affected the growth of cities in other parts of India, but this section has illustrated the particular effects in Punjab. Nevertheless, to assess the growth of cities within the Punjab, an understanding of the particular social and political history of the region is required.

Punjab is a region which tends to be associated with the agricultural sector rather than the industrial sector and with rural rather than urban living. It is often referred to as the 'breadbasket' of India because of its agricultural successes. Though it is true that Punjab's high levels of economic output are primarily due to the Green Revolution and the commercialisation of agriculture, it is inadequate to view this aspect of the economy in a vacuum. It has been posed by some people that the Green Revolution in Punjab has had detrimental effects on the social and class structure and has been unequal and differential in its effects. The rise of large landowners with access to capital and the further marginalisation of poor, labouring groups bears witness to this view (Frankel 1971; Byres 1983). Where industrial growth has taken place this has been a result of the work of individual entrepreneurial effort, such as in Ludhiana, rather than the intervention of the state. However, agricultural growth has not been adequately matched with industrialisation which has meant that surplus labour from an increasingly capital-intensive sector has migrated to cities for employment (Oberai and Singh 1983). The effects of this have often resulted in high rates of unemployment and a lack of sufficient housing, most adversely affecting conditions of urban poverty. In Amritsar the static nature of the position of the urban poor is reflective of the pattern of economic development, a point which will be taken up in subsequent sections in this chapter. Green revolution-style agricultural development and a steadily declining labour-intensive textile trade has led to an economy which has both failed to provide employment security to the poor while also failing to generate a dynamic housing sector to fulfil the accommodation and service needs of the poor of the city. This is despite the higher levels of per capita income that is enjoyed in the Punjab compared to other states and the relatively healthy resources of the state. This type of development in turn has contributed to the complex ethnic politics of the region which has both attracted migrant labour from other regions and aroused anti-centre sentiments.

Ethnicity and Communal Politics

The city of Amritsar has historically been a place where different religious communities have coexisted. The region of Punjab had, prior to partition, represented a diverse cultural, literary and religious agglomeration. 'Punjabi identity' has been attributed to geographical, socio-political, administrative, poetic, linguistic, cultural and historical factors which account for a regional identity based on a multi-dimensional identity.[8] Ethnic and communal identity politics in the region have been highlighted at a number of points within the social and political history of Punjab. The partition of Punjab, as was discussed earlier, created divisions between communities which had geographic, demographic and political implications to which the repercussions are still being felt today. The physical separation of Punjab into Indian and Pakistani Punjab in 1947 meant that religious identities became dominant to a secular Punjabi identity. Though the city of Amritsar had been ruled by Ranjit Singh and had been created largely due to the building of the Golden Temple, the population of Amritsar before partition revealed a Muslim majority (See Table 3.3). Partition resulted in the overwhelming majority of Muslim residents leaving for Pakistan and in an increased demographic representation of a Hindu and Sikh majority.

The Punjabi Suba Movement was another momentous period within the history of Punjab. At the forefront of the movement were disaffected Sikh leaders who argued that the Sikh community had not gained from the partition of the land in 1947. While the movement was, in name, based upon linguistic recognition, it represented the communal demands of a particular religious community, the Sikhs. The link between language and communal politics came to the fore in the 1961 Census of India when many Punjabi-speaking Hindus declared Hindi as their spoken language thus creating a further wedge between the two religious communities (Tully and Jacob 1985). It resulted in the reorganisation of the state's boundaries in 1966 on the basis of language whereby the state of Haryana was carved out as the Hindi-speaking majority area.[9] However, the Punjabi Suba movement was not isolated from other class, religious and linguistic identities. The Arya Samaj movement emerged as the voice of urban middle class Hindus who, in Punjab, sought to align themselves with a pan-Indian, Hindu identity rather than a secular Punjabi one. The Arya Samaj similarly made a considerable contribution to the communalisation of consciousness and in the erosion of a secular regional identity (Banga 1996). These two movements together resulted in a

59

heightened tension between the Sikh and Hindu religious communities in Punjab which has remained as a dominant factor in contemporary Punjab politics.

Another important period in Punjab's history which deeply affected ethnic and communal politics was Operation Bluestar, the army code name for the military storming of the holiest Sikh shrine, the Golden Temple. In 1984 the Golden Temple (*Harimandir Sahib*), was attacked by the Indian army upon the orders given by the then Prime Minister Indira Gandhi. The occupation of the Golden Temple by militants demanding the establishment of a free and independent Khalistan[10] had threatened the central government's authority in the region thereby resulting in an offensive by the central government authorities. The demand for Khalistan is based upon political, religious as well as economic arguments. Grievances around the underdevelopment of Punjab's industrial sector in comparison to other states has been one of the primary points raised by Khalistani activists (Gill 1999). The economic viability of an independent Khalistan from the Indian union, given the region's economic output through agriculture, has been particularly highlighted. The agricultural successes of Punjab, as is argued, have not been rewarded to the state itself but in subsidising the development of other regions. Amritsar, as home to the Golden Temple and Sikh organisation and leadership, has acted as the centre of the political movement since 1984 despite the rural base of the militant activities.

Operation Bluestar was perceived, by not only the local Sikh community in Punjab but also the global Sikh community,[11] as an attack upon their religious rights and security and marked the beginning of a period of strained centre-state relations (Oberoi 1987). Several months after the military attack, Indira Gandhi was assassinated by her two Sikh bodyguards. The anti-Sikh riots which occurred immediately thereafter resulted in the murdering of approximately 5000 Sikhs in Delhi alone and in a growing politicisation of Sikh identity in relation to the nation of India.[12] Tensions between the Hindu and Sikh communities which had begun after the 1961 Census and the 1966 reorganisation of states were now further solidified through an emerging dichotomy of religious affiliations and distrust between communities.

Caste has continued to play a role in contemporary Punjab's ethnic and communal make-up. The Mandal Commission in 1990 invoked protest on a national level against the reservation of government jobs for Scheduled Castes. Higher castes, reluctant to make concessions for a group much larger than themselves, were successful in ousting the Janata Dal government which had introduced the bill. In Amritsar resentment of the reservations in schools and colleges for scheduled castes still remains a point a conflict. Simultaneous to the above mentioned Hindu-Sikh communal history, the position of lower castes within the social and political environment has continued to reflect their marginal position within all forms of social organisation. Religious communities have been more directly divided along lines defined by macro-political processes while caste groupings have aligned themselves within a complex set of criteria. For example, the Akali Dal even before the political turmoil of the 1980s had emerged as the party of the Sikhs and traditionally relied upon support from the Jats.[13] However, Majbi Sikhs have not been a vote bank for the Akali Dal, despite being a sizeable Sikh community and have instead supported Congress or leftist parties due to caste and class animosities towards their employers, the Jat farmers (Kohli 1990).[14]

Generally, Scheduled Castes of the various religions extending beyond the Hindu-Sikh dichotomy have remained at the margins of Punjab's political activity. Political parties such as the Bharatiya Janata Party (BJP) and Congress have utilised the economic insecurity of these communities in order to attract votes and political support. Recently, however, the Bahujan Samaj Party (BSP) which represents the lower castes and has directly confronted the issue of reservations has gained popularity among Amritsar's lower castes. As discussed earlier in this chapter, the influx of economic migrants from Bihar and Uttar Pradesh to Punjab has reflected a change in the labour structure of Punjab. An overwhelming majority of such migrants belong to the Scheduled Castes and have also begun to show an interest in the political environment of Punjab. This will no doubt have far-reaching long-term effects not only upon electoral politics in the state but also upon how these migrant communities will wield their voting power in attaining patronage from local political interests.

This chapter has given a cursory overview of the main themes in Amritsar's development without going into too much detail. The historical background of Amritsar reveals a complex history of growth, development and decline. The selection of the site for the construction of the Golden Temple in the sixteenth century began the period of growth of the city with the establishment of *katras* which were further enhanced by the reign of Maharaja Ranjit Singh who consolidated not only the city of Amritsar but also the region of Punjab up until his death in 1839. The erection of the wall by Ranjit Singh against threats of attack from Mughal and Afghan invaders left its mark upon the city by containing the spatial expansion of the city with a densely populated walled city core. The annexation of Punjab in 1849 by the British began a century of structural changes to the spatial and administrative structure of the city of Amritsar. The construction of the cantonment and civil lines demarcated the spatial division between native and British localities, a legacy which has been continued after independence between the elites and the lower and middle class groups.

The partition of Punjab in 1947 marked the most significant period of change to the city of Amritsar. The exodus of forced inward and outward migration caused the city's population and economic significance to decline, creating imbalances in the labour structure, demographic make-up and economic activities. The position of the poor in post-partition Amritsar has been largely effected by the city's declining role in the region. The number of employment opportunities available to the poor is limited to predominantly informal, irregular sources of income. Political factors have also contributed to the demise of Amritsar as an urban centre. Most recently, the storming of the Golden Temple in 1984 resulted in over a decade of state repression and anti-state violence which has further entrenched the city into a pool of stagnation. The overriding picture which emerges from this history is that the poor of Amritsar today occupy a structurally marginal position within a city which has itself occupied a structurally marginal position to economic and development processes. Their struggles and obstacles in finding shelter for themselves has been marked by this history.

---

[1] See Ian Talbot (1994) "State, Society and Identity: The British Punjab, 1875-1937" in Singh and Talbot (eds.) *Punjabi Identity: Continuity and Change*, Manohar for a more in-depth analysis of the effects that colonialism had upon religious identities in Punjab.

[2] See Virinder S. Kalra and Navtej K. Purewal (1999) 'The Strut of the Peacock: Travel, Partition and the Indo-Pak Border' in Raminder Kaur and John Hutnyk (eds.) *Travel Worlds: Journeys in Contemporary Cultural Politics*, Zed: London.

[3] Census of India Final Population Totals (1991).

[4] Class I cities are defined as those cities exceeding populations of 100,000.

[5] See Byres 1981 for a comprehensive analysis of the effects of the Green Revolution upon the agrarian structure.

[6] India's national development strategy has been governed by five-year national plans with the first beginning in 1951. For a detailed reflection on the experience of the past forty years of development planning in India see Terence J. Byres (1994) (ed.) *The State and Development Planning in India*, Oxford University Press: Delhi.

[7] See Chapter Four.

[8] See Singh and Talbot (1996) (eds.).

[9] The independence of Bangladesh from Pakistan in 1972 was also a response to the growing significance of language as a basis for regional identities and political formations.

[10] The demand for Khalistan is not a recent movement. It has its roots in the period prior to independence when the All India Muslim League had begun deliberating the possibility of a Muslim state, Pakistan. Khalistan, as argued by Dr. V.S. Bhatti in 1940, would be a Sikh state which would act as a buffer between two unfriendly neighbour states, India and Pakistan. See Sukhmani Riar (1996) 'Khalistan: The Origins of the Demand and Its Pursuit Prior to Independence, 1940-45,' Singh and Thandi (eds.) *Globalisation and the Region: Explorations in Punjabi Identity*.

[11] For a global perspective on Punjab and Sikh studies, see Pritam Singh and Shinder Singh Thandi (eds.) (1999) *Punjabi Identity in a Global Context*, Oxford University Press.

[12] Thandi (1994) notes that the death toll in Punjab during the period 1981-1993 may never be known but that estimates show between 20,000 and 30,000 to have died and between 20,000 and 45,000 'involuntary disappearances'.

[13] The Jats of Punjab have been an active force in contemporary Punjabi politics. Jats are a farming caste community. See Joyce Pettigrew (1975) *Robber Noblemen: A Study of the Political System of the Sikh Jats*, Clarendon Press.

[14] Khushwant Singh (1966) explores the historical background to the Akali Dal party and its ascension within Punjabi politics in *A History of the Sikhs, Vol. II: 1839-1964*, Princeton University Press.

4   The Research Method

Introduction

In further exploring some of the ideological and theoretical positions introduced in Chapters One and Two, the research method of this study takes into account both practical and ethical issues. The 'slums and squatter settlements' stereotype has a number of connotations when one examines the multiplicity of legal situations, tenure arrangements and economic positions of settlements where the poor find shelter and therefore renders a more critical examination of definitions and meanings. The diversity of definitions and terminology used within the field of poverty and shelter reveal some of the incessant controversies that exist, for example, between legalistic frameworks and more decentralised *laissez-faire* self-help policies. Such issues around labelling and terminology will be considered in this chapter. Furthermore, a typology designed specifically for this study of Amritsar will also be developed here which will form the basis for the remaining chapters and their analyses.

In the remaining sections of the chapter I will present the methods of research employed in the study and will reflect upon the fieldwork experience. This will highlight the way in which the research was conducted in identifying some of the socio-economic and spatial aspects of shelter access while also acknowledging some of the limitations of the study. I conclude with a discussion of my own position as a researcher from the West and some of the ethical issues involved in doing research and fieldwork in a developing country context.

Definitions and Terminology

Many perceptions exist as to what position the urban poor in Third World cities occupy and how, if at all possible, to define and identify poor urban communities. Much official information of both international agencies and Third World governments use poverty line statistics in order to specify the definition of low-income. In South Asia a high proportion of urban populations are said to be living below the poverty line with 60 percent in

Calcutta, 50 percent in Madras, 45 percent in Bombay and 45 percent in Karachi (Cheema 1986: 2). While a large percentage of total urban inhabitants in these cities are living in slums and squatter areas, the concept of a homogenous 'low-income' or 'poor' community is not unquestionably upheld in this study. Rather, a terminology which interchangeably combines low-incomes, relative poverty and substandard housing conditions is loosely applied. Studies on urban poverty have attempted to identify the urban poor both in terms of economic status, most commonly through the establishment of poverty lines, and social characteristics such as migration history and average household size.

The popularly cited Human Development Index (HDI) published by the United Nations Development Programme (UNDP) administers a system whereby countries are ranked according to the consideration of a number of criteria including per capita income, gender equality, nutrition, health, education (UNDP 1996). However, the use of purely statistical indicators has been criticised for its simplification of what are considered to be a complex set of development issues which do not take into account political and social change. The concept of poverty in India has been defined by a national study of economic, demographic and shelter characteristics of the urban poor in India through a survey of twenty urban centres by the National Institute of Urban Affairs (1989) established the poverty level as those households earning Rs. 122 per capita per month or Rs. 7300 per household per annum. The Government of Punjab in 1995 in its Green Card scheme[1] marks Rs. 18,000 as the highest yearly household income for those to qualify for the scheme. Another report by the National Institute of Urban Affairs in its *Handbook of Urban Statistics* defines the poverty line by "the income level below which a minimum nutritionally adequate diet plus essential non-food requirement are not affordable" (1993:110).[2]

Authors researching poverty in developing and developed countries have come to a number of conclusions on the use of definitions in analysing poverty. A study by Lewis (1966) established the term 'culture of poverty' to explain for the existence of widespread poverty through a notion that poor people are poor because they are living in a culture of poverty which encourages them to accept their condition as fate. They therefore lack aspirations or incentives to improve their lot and are trapped in a cycle of poverty. Desai and Pillai (1990) similarly describe slums as areas of darkness and despair. Turner (1967, 1969) and Mangin (1967) opposed this approach in their work in Peru where they observed that the poor, if given uninhibiting circumstances, could find innovative ways out of their poverty and build for themselves adequate, serviced housing. Gilbert (1982), however, states that local factors need to be assessed before making comparisons of conditions of the poor in different Third World contexts so as not to fall into the trap of forming subjective and ethnocentric methods of analysis. The rejection of the 'culture of poverty' approach has been particularly vocalised by the self-help school. The argument that the poor occupy a place within a sub-society is strongly opposed. The self-help literature sees the aspirations of the poor as similar to those of other non-poor communities and that, given the opportunities, the poor will improve their situation and fulfil their aspirations (Perlman 1976).

The political economy or structuralist approach, however, views poverty as a result of structural conditions which operate to the disadvantage of the poor. In order to continue sustaining the generation of profit-maximising processes within the economy, poor populations are necessary to provide cheap labour and services. Harvey's (1988) analysis sees labour through its relationship to capital. While capital is mobile, labour is inert meaning that when people move, they only do so in relation to the movement of capital (Dickens 1990). Economic processes and social relations of production are the determining factors of poverty (Castells 1978). The reproduction of a workforce which can participate in labour is necessary for capital to remain productive and profitable. Therefore, poverty and the sustenance of a poor population is, to a certain degree, a requirement of capitalist societies. In the case of India, however, the 'mixed economy' complicates the analysis of the persistence of poverty. Until the early 1990s the Nehruvian model of centralised economic planning and private sector economic development operated simultaneously.

Following on from the political economy analysis of poverty, the central premise of this study is that the poor are structurally marginal to the economic processes and social relations in Amritsar as opposed to Lewis' 'culture of poverty' thesis. This study has targeted households which are residing in settlements of predominantly substandard physical conditions and infrastructure and therefore assumes that those people living within such settlements cannot afford to live under better conditions and have incomes which are insufficient in affording better housing. The majority of urban poor households can only gain access to extra-legal forms of housing, resulting in the bulk of the low-income housing literature emphasising illegality. The evolving nature of the unregulated housing markets has resulted in middle income groups also being excluded from formal means of housing access. Therefore, low-income is used throughout the book not exclusively as a definition according to the level of income, but also in the inability of households to purchase or rent legal, structurally solid and adequately serviced housing.

While low-income housing settlements are defined here within the typology to be developed later in this chapter, there is a tendency within the literature to use 'slum' to describe all residential locations where the urban poor are located which is worth briefly examining. An early report by the United Nations of urban land policies defined a slum as:

> ...a building, group of buildings, or area characterized by overcrowding, deterioration, unsanitary conditions or absence of facilities or amenities which, because of these conditions or any of them, endanger the health, safety or moral of its inhabitants or the community (United Nations Secretariat 1952: 200).

Slum localities have officially been defined by the Indian government as "areas where buildings are unfit for human habitation by reason of dilapidation, overcrowding, faulty arrangement and design or buildings, narrowness of streets, lack of ventilation and light" (Government of India 1981: 29; Crook 1993: 134). The National Institute of Urban Affairs (1993: 54-55) also uses the term 'slum' in its estimate of the percentage of slum population in Indian states (Table 4.1).

Table 4.1 Slum Population in India and Selected States, 1981 and 1990

	% of slum population 1981	Estimated urban population 1990 (1000)	Estimated % slum population 1990
India	17.47	2415.44	21.2
Bihar	37.50	137.72	23.7
Gujarat	14.45	155.05	20.0
Himachal Pradesh	23.31	4.58	20.1
Jammu and Kashmir	49.76	19.44	32.3
Maharashtra	19.62	312.55	20.0
Punjab	25.11	25.11	20.0
West Bengal	20.96	198.57	25.0

Source: The National Institute of Urban Affairs (1993: 54-55).

In 1975 the Municipal Committee in Amritsar on the basis of the Punjab

Slum Areas (Improvement and Clearance) Act 1975 defined slums as those areas where:

> ... the competent authority upon report from any of its officers or other information in its possession is satisfied in respect of any area that the buildings in that area are in any respect unfit for human habitation or are by any reason of dilapidation, overcrowding, faulty arrangements and design of such buildings, narrowness or faulty arrangement of streets, lack of ventilation, light or sanitation facilities or any combination of these factors detrimental to safety, health and morals.

The broadness of this definition means that inner-city tenements, squatter settlements, quasi-legal settlements, pavement dwellers as well as middle income areas where infrastructure is lacking could all come within this single category.[3]

An underlying issue which arose during the sample selection was the usage of terminology to describe areas occupied by low-income communities with sub-standard conditions. In the literature a variety of terms are deployed.[4] 'Spontaneous settlements' has been used to describe the need-based inspiration of user-builders to construct houses (Patton 1988) while 'illegal settlements' has been used to describe the legal-illegal dichotomy (Grimes 1976); 'informal settlements' and 'unregulated markets' have been used to describe the dual-economy approach which acknowledges the informal sector as an active supplier of many needs, of which housing is one, to the poor (Payne 1989; Shakur 1993); Sarin (1989) uses the term 'non-plan settlements' to describe the development of settlements outside of Western-oriented urban planning models. 'Low-income housing' is also a common term in the literature which signifies income as a main factor (Payne 1984; Rakodi 1995).

A majority of the literature on India and other Asian countries uses the terms 'slum' and 'squatter settlement,' all-encompassing descriptions for a range of different types of housing for the poor (Mitra 1994; Desai and Pillai 1991; Das Gupta 1984). A comparative study of Colombo and Bangkok uses the term to describe gender relations and slum culture in two low-income settlements (Thorbek 1994). Thorbek uses the term 'slum culture' to describe the interactive social relationship that exists between migrants and the indigenous poor in slum areas. In another study Desai and Pillai (1991) classify the slums of Bombay into three types: single multi-storeyed buildings, semi-permanent legal and illegal structures and hutment or squatter settlements known as *zoppadpattis*. Thus, as is exhibited by these studies, much of the literature on low-income housing in

South Asia, while also utilising other more specific terms such as *jhuggies*, squatter settlements, *bastees* and *katchi abadis*, relies heavily upon the more general term 'slum.'

Critical analysis of the relevance of the term is also evident in the literature. The evolving and increasingly complex role of poor residential areas has led several authors to note the need to cautiously apply definitions where they may or may not be appropriate. Crook (1993) argues that the romantic, rustic image of rural villages disallows the term 'slum' to be used in reference to unserviced, substandard constructions in small non-industrial towns and villages. He stresses the continuity that exists between rural and urban built environments and that the stereotypical slums of highly urbanised cities are not necessarily relevant in smaller urban settings. Desai (1995) recognises the openness of the term in her examination of Bombay's more dynamic slums and notes that the term 'slum' has been perceived in a number of ways depending upon the observer's point of reference. Poor migrant households might view it as a positive sense of security while the middle and upper classes as well as urban planners view it as a negative component of urban development. Thus, it can be argued that the generalised term 'slum' which has been commonly used in studies of low-income housing in India is highly interpretable and therefore adaptable to the specific urban context to which it is applied. Here, the term 'slum' will be utilised only in reference to other comparative studies and not applied to the context of Amritsar. In the following section, a typology designed specifically for the Amritsar context will be developed in an attempt to provide a sufficient terminology in describing and analysing poor residential settlements in the city.

Selecting a Sample

A major obstacle in researching urban poor communities is that official statistics do not often include them. Information on low-income settlements is more readily available from community residents themselves than through official documentation. Even as the scope and span of research continues to extend to new regions and cities in developing countries, the data bank on low-income communities is still far from sufficient. Generally, official information that is available tends to be unreliable, and at times dubious, and not particularly representative of the urban poor who often do not count in the figures.

Except through estimated figures, there is no way of knowing numbers of pavement dwellers, squatter settlers, and other low-income communities (Hardoy and Satterthwaite 1989; Gilbert and Gugler 1982). Therefore, the combination of qualitative and quantitative methodological tools, given the obstacles of data unreliability and inadequate resources, are essential to the generation of research that focuses on the urban poor and the realities of their living and working conditions (Brewer and Hunter 1989). Such techniques can also help to construct a project which more adequately features elements of marginalised, often unrepresented and misrepresented, groups such as slum and squatter settlement residents (Ragin 1994).

The lack of existing information on poor communities in the city due to the above mentioned reasons required a social survey for this study. The selection of the sample was a process of exploring the city of Amritsar, identifying low-income communities and neighbourhoods, classifying these low-income areas, and eventually selecting a sample from this classification. The identification stage involved reconciling official lists of notified slums with unlisted slums and *jhuggies*. Some settlements which were listed on the Amritsar Municipal Corporation list were occupied by middle-upper income families where basic facilities were present. These areas were excluded while areas which were not 'notified,' generally for political reasons, were added to my own list and tended to be the worst-off in terms of housing standards and facilities provided.

The sample was selected from the total Amritsar population as a percentage of the low-income settlement population. With Amritsar's total population at 709,456 (Census of India 1991) and an approximation of 422,900 as the total low-income population in the city, the area around the walled city, which is the largest zonal area in terms of low-income population, was chosen as the focus of this study. An informal list was compiled of all settlements which were identified as being low-income, unserviced areas. This became the framework for consideration in the study. The criteria in determining this framework were spatial location, physical conditions, quality of urban services, legality, form of development and socio-economic characteristics. These criteria were selected for identifying the sample as they provided information about the geographic location of settlements within the city as well as their social, physical and legal characteristics.

The city was divided into four Zones. Zone One was defined by the area inside the walled city with a low-income population of 50, 000. Zone Two included the areas surrounding the walled city and the outer circular road with an estimated low-income population of 191, 400. Zone Three included the areas within the urban boundaries of Amritsar which extended beyond the immediate vicinity of the city centre and has an approximate low-income population of 39, 000. Finally, Zone Four was defined as the rural areas which fall within the Amritsar Municipal Corporation boundaries but which are on the outskirts of the city. This area was estimated to have a low-income population of 142, 500. Figure 4.1 shows the demarcation of the four zones.

Figure 4.1 Zonal Map of Amritsar

0   3kms

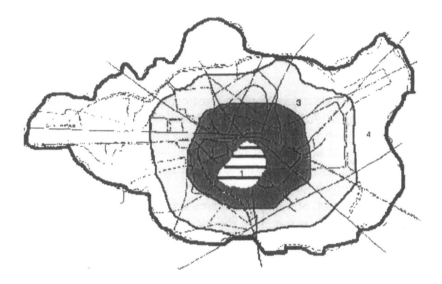

The walled city's total population is approximately 250,000 out of a total population of approximately 709,456 so it is necessary to clarify why it was not directly included into the sample of study. The walled city consists of Amritsar's oldest housing stock as the inner-city core. The walled city has been at the centre of the evolution and expansion of the city, urban morphological changes and the social movement of communities. However, the residential use of the walled city is increasingly being substituted by commercial and even industrial purposes (Batra 1985). This has resulted in a complex dynamic of increased commercial values of properties in the walled city for business purposes and in the physical

71

deterioration of older residential areas. Many middle and high income groups, however, maintain their ancestral homes in the walled city as they have shifted their residential base to the new posh localities developing around Amritsar. Low-income groups, in the meantime, had been excluded from the walled city's residential market as the demand for space had increased during Amritsar's growth period. The presence of low-income groups around the walled city represents the significance of squatting to Amritsar's poor communities who are only recently regaining access to the walled city's housing stock through rental markets. In the case of this study, it was therefore appropriate to focus on the area around the walled city as it has historically provided the poor with housing. The area also contains the highest proportion of illegal, unserviced housing and is subject to the jurisdiction of a number of different public agencies.

Housing Structures

The types of housing structures that are most commonly occupied by low-income households in Amritsar can be generally described within four categories:[5] *jhuggies, kaccha* houses, *semi-pacca* houses and *pacca* houses. Each of these types of housing fall within the basic housing system of the city and carry with them implications of the condition of security and investment of the household.

The construction of a *jhuggi* requires the least amount of financial and time investment and is therefore the most viable form of housing for households with little or no security from eviction. The most common materials used in Amritsar are wood planks and cane for the walls, tin and metal for the structure, and plastic sheeting and jute for the roof. The construction of a *jhuggi* requires a small fraction of the cost of building a more solid house. However, while cost is a main reason for poor families to build *jhuggies*, the common characteristic among *jhuggi* households in Amritsar is the nature of their tenure status. Most *jhuggies* either exist on outer fringe areas of the city on commercially undesirable land or on squatter settlements within the city which are on highly disputed plots of public land. The physical development of these settlements is therefore negatively affected by the insecurity of tenure (van der Linden 1983). Those living in *jhuggies* are mostly economic migrants from the eastern states of Bihar, Uttar Pradesh and West Bengal.

Kaccha houses are generally self-built structures made from bricks plastered with a mud mixture. This type of structure requires a high level

72

of maintenance, particularly due to rain, and is more appropriate for rural areas where the irregularity of agricultural work allows for time to be spent on the maintenance of the house (van der Harst 1983). In areas which are becoming more urbanised, the accessibility to mud becomes an obstacle. The costs of bricks and mud are only marginally higher than the materials used to build *jhuggies*. However, the long amount of time required for construction and regular maintenance reflect the demands on family labour efforts. Nonetheless, the fact that there is an on-going investment into the house also means that the household feels a sense of security to do so.

The *semi-pacca* house can take the form of cement blocks, bricks and plaster. The use of plaster and cement blocks makes the structure more durable compared to the *kaccha* structure. The plastering of walls increases the house's resistance to the effects of rain and increases the life of the structure. The construction of a *semi-pacca* house is rarely done by the household alone. The use of skilled labour is most common, while the household members often contribute to the building process in order to save on labour costs. The *semi-pacca* house is considerably more expensive than the *jhuggi* or *kaccha* house. In his study of Karachi slums, van der Harst estimates a *semi-pacca* house to be seven times more costly than a *jhuggi* or mudhouse of the same size (1983: 63). For this reason, much of the household savings that go towards housing is oftentimes lost in the maintenance of *kaccha* structures before an investment into the initial costs of bricks, cement and plaster can be afforded for a *semi-pacca* house.

The *pacca* house is similar to the *semi-pacca* structure with the significant addition of a solid roof and foundation. The roof is reinforced with concrete in order to hold a second storey. The foundation is made particularly secure in order to support the weight of a second story. This requires the expertise of a building contractor who has both the knowledge of necessary building materials and the capital in order to carry out the task. This type of house is unusual in Amritsar's low-income self-help settlements due to the high initial expenditure required for construction and materials. *Pacca* structures are generally only found in middle and upper income private settlements and in government housing sites. Indeed, this is a main feature of the appeal of government housing to low-income groups. Investment into building and maintenance costs of an equal standard house in a self-help settlement is unaffordable for low-income residents. Without the savings to invest in *pacca* private self-help housing, government housing for its structural advantages is a housing option which is in high demand.

Constructing a Typology

Studies of South Asian cities have constructed a range of typologies on the basis of various considerations applicable to their particular urban contexts. Sarin (1982) categorises the poor residential settlements in Chandigarh as planned and non-plan settlements which is derivative of the city's particular context as a planned post-colonial city. Sandhu (1989), in his study of Amritsar's 'slums,' categorises low-income settlements into two types based on geographical location: those within and just outside the walled city and those on the periphery of the city. In study of land supply in the slums of Calcutta,[6] Roy (1983) reveals a detailed eight-tiered typology which reflects upon historical development, land tenure as well as socio-economic make-up of residents: conventional *bustees*, legal refugee colonies, extra-legal refugee colonies, squatter settlements, jute lines, private self-help housing, old rented walk-ups and government tenements for low-income people. While the geographical base of Sandhu's categorisation is utilised and appropriated in the spatial selection of survey settlements for this study (See Figure 4.1), the typology developed here more closely resembles Roy's observations of Calcutta's slums in its method of combining historical, tenure and socio-economic processes. Three types of settlements were identified as housing types which were being accessed by the poor and which form the typology designed for Amritsar: private self-help settlements, state-assisted sites and *jhuggi* colonies (Table 4.2). The diversity of systems required a classification which was adequate as a description, yet open enough to include the different types of housing access and accessibility in Amritsar. As the three-tier typology will be continually referred to in the this study, I will give a brief descriptive introduction of each type of settlement which will be further expanded upon in the remaining chapters.

Table 4.2 Typology of Low-income Settlements in Amritsar

Type	*Jhuggies*	Private Self-Help	State-Assisted
Structure	*kaccha*/tent	*kaccha/semi-pacca*	*semi-pacca/pacca*
Status	illegal	quasi-legal	legal
System	marginal	informal	government *ad hoc*

Jhuggies exist on either marginal, temple or disputed land where insecurity of tenure is detrimental to the physical development of the housing structures (Schoorl et al 1983). Harassment by police or other interests in their removal is constant and is particularly evident in the make-shift tent and *kaccha* structures which, for the resident-builders, require the minimum amount of immediate investment. Services are non-existent except for a tap or hand pump for water which services the entire settlement. Often, it is this source of water, in addition to the availability of free land, that draws people to these areas to make their homes. In Amritsar, temple land is also a source of land for low-income settlement. These are places where water is readily available and where there is generally no commercial development to compete with for land. The Durgiana Mandir, a religious tourist site, is a common 'first stop' for new migrants from Bihar and West Bengal, and on several occasions residents have been forcibly relocated to land less visible and less desirable. Possibly the largest sense of security for residents comes from the fact that *jhuggi* settlements, are less heterogeneous in character than other types of residential areas where the settled communities generally come from similar regional, caste and ethnic backgrounds.

   Private Self-help housing is the most prevalent type of housing within the low-income settlements of Amritsar. Some settlements were built-up areas prior to partition which were then occupied by local residents. Other private self-help settlements were established through the influx of partition refugees who were left with the only option of squatting upon vacant land but whose current legal status is dependent upon the government's ability or willingness to award legal titles. Land has been obtained by the poor through non-commercial processes on public land. A vast area of dumping ground outside the walled city has been settled upon and developed by residents themselves. Water and sanitation are two main problems faced by these settlements. Contact with local political parties *vis-à-vis* votes in upcoming elections is extensive and is represented in the installation of communal taps and hand pumps and in some cases in the granting of registries. There is increasing evidence of reselling of land by original inhabitants to private buyers. This is due to the recent moves by the local authority of regularising settlements through the granting of registries.

   Public Housing for low-income groups (LIG's) and Municipal Corporation Housing both fall within the category of *state-assisted settlements*. Municipal Corporation tenements were constructed for specific target groups: scheduled castes, namely people working as sweepers in government buildings, and people displaced as a result of floods that ruined agricultural land and the war between India and Pakistan that affected rural border areas. These tenements were constructed to an utmost basic level with no facilities, no drainage, and only minimal housing materials used. Few original allotees actually live in these tenements as the construction process was not followed up by any official allotment process. However, the original inhabitants are still living in these flats and there is no sign of reselling, most likely because of the lack of registry.

Public housing schemes are still being constructed to target low-income families. These schemes are generally located on land distant from the main commercial areas and places of employment because of the relatively low market value. Allotments are made through a tender system whereby people submit applications for houses or flats. Those granted allotments are then given registry upon fulfilling the monthly instalments set over a period of time. The instalment system is what makes this form of housing attractive as the access and availability of cash is the biggest obstacle to housing investment by the poor. As has been the case with sites and services and conventional schemes in other Third World cities, the location of these schemes on marginal land has meant that many cannot afford the travel expenses and time to reach places of employment. Even though services are available, they are not maintained to very high standards. However, because security of tenure through the official nature of the allotment system and the granting of registries has been a positive aspect of this form of housing, middle income families have also sought to benefit from such schemes. The incidence of renting is also significantly higher in such schemes than in private self-help settlements, as will be pointed out in Chapter Seven.

The Survey

Traditionally, surveys and anthropological methods have been viewed as oppositional research strategies where objectivity and subjectivity have been often seen as having conflicting results (Whyte 1964; Whyte and Alberti 1993). Each method offers a different type of information which cannot be adequately compared with the other. This research has therefore attempted to combine both qualitative and quantitative research methods in order to bring into the analysis both quantifiable elements such as income, population figures and housing structure as well as qualitative elements such as attitudinal data and interpretative analysis. In-depth interviews have been used in the research to enhance and supplement the findings of the survey. It is this combination of 'hard and soft data' (Bulmer and Warwick, Marsh 1982) which has been applied in the research to offer insights of the data collected in the survey.

The fieldwork in Amritsar for this study was conducted in 1995. In total 275 households were interviewed in 15 settlements which represented approximately 24 percent of the total low-income settlements in Zone 2. Within this division, the typology representation of selected sample settlements was further calculated at 18 *jhuggi* settlements, 186 self-help settlements and 71 government-aided settlements. The sample survey in this study was selected from the typology, and the spatial location of the settlements surveyed in this study are shown in Figure 4.2.

Figure 4.2 Spatial Location of Settlements by Typology

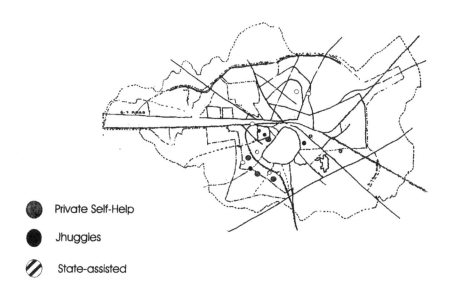

● Private Self-Help

● Jhuggies

◉ State-assisted

Table 4.3 Construction of the Sample Survey

Category of Sample	*Jhuggi* Settlements	Private Self-Help Settlements	State-Assisted Settlements	Total
Number of Households Interviewed	18	186	71	275

The need for primary data required the use of a detailed questionnaire. Amritsar being a provincial town, available information on the city and its low-income residents was limited. A number of thematic areas were selected in order to obtain information regarding social, economic and housing characteristics of each interviewed household. The purpose of the survey was to answer the research questions as laid out in the research design and also to generate a database of the low-income housing settlements in Amritsar utilising the typology as shown in Figure 4.3.

As the primary objective of the survey was to assess social access to shelter, eight main categories were outlined which would provide the necessary information from which to draw conclusions about the patterns of social access: ethnic and caste background, household composition, income generation, household finance, dwelling information i.e. structural aspects and services, opinions about house and neighbourhood, process of acquisition of house and migration. The questionnaire was designed to acquire a basic level of information from all respondents but also to encourage more detailed accounts where respondents were willing to do so. The questions were loosely posed so as to encourage respondents to provide more detailed responses. However, while some chose to give this information fully, others were more reluctant and did not. The final section on the nature of occupation, tenure and legality was given to owners and renters separately, and a separate section was designed especially for *jhuggi* residents for whom issues of security and occupation required more specific details and questions of ownership were irrelevant.

Two research assistants, both students at the Guru Nanak Dev University Department of Planning, conducted the interviews while I oversaw the interviews and made further notes and observations. Interviews were conducted in Punjabi and Hindi and were written down in English. Only one person was interviewed in each household irrespective of the head of household. This was done as an attempt to avoid a gender bias in the sample. However, despite these efforts, a gender bias in the survey still persisted. As a research team of two males and one female, we were perceived by many residents as potentially influential people with possible connections to urban and housing policy improvements in their locality. As a result, when men were at home during our visits, they stepped forward as the 'obvious' family member to answer our questions. Even when the research team insisted that the woman take part in the interview, her responses were often not accepted by both male and female household members as representative of the household. This proved to be a limitation of the use of a questionnaire.

There are, however, a number of other problems in the use of questionnaires in surveys. The questionnaire serves the purpose of streamlining questions and responses in line with the objectives of the

Figure 4.3 Distribution of Respondents in the Typology

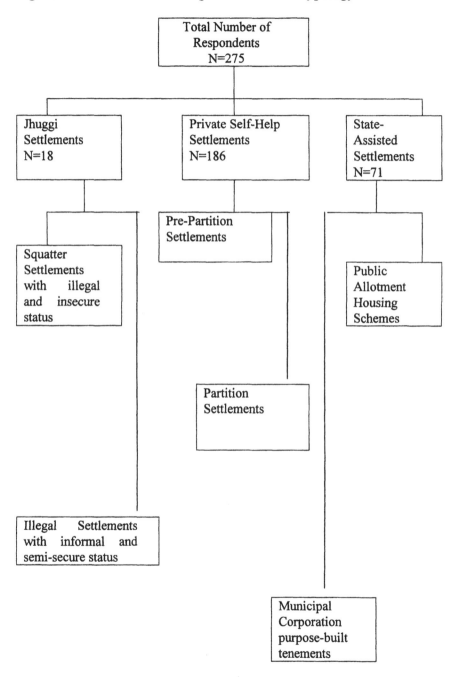

survey. Survey research relies upon variables to define a set number of values for each case (Marsh 1982). In social research the translation of opinions and attitudinal information into finite sets of values is therefore a highly constructed process. This 'variable language' requires a fixed coding system in which all responses can be applied. This presents limitations to the applicability of subjective responses and unexpected responses which often do not fit within the established value categories. While many questions were presented with open-ended as well as closed options for response, the tendency of many respondents was to provide the closed option where they were unsure of the validity of their response. It is due to this limitation of quantitative methods that qualitative methods through case studies and observations were also employed in the study in order to adequately represent the subjective and attitudinal elements of the responses.

Case studies were selected from the sample survey respondents. They were selected on the basis of the overall settlement profile and the respondents' period of stay in the settlements. Questions asked went into more detail about information obtained about moving to their house, reasons for selection of residence and the social and political dynamics of the settlement. Since interviews were semi-structured, respondents were encouraged to discuss the aspects of their housing which they felt were important which allowed for them to steer the interview's direction.

Case studies provided valuable information not only about personal experiences, but also of settlement histories. It was through these interviews that a fuller understanding was developed of the dynamics of the housing systems in Amritsar. The respondents' concerns, aspirations and observations oftentimes elicited information which had not been explicitly laid out in the questionnaire, in particular with regards to the political dynamics within the settlement and in relation to how housing is accessed. The interviews were translated from Punjabi and Hindi to English.

Ethical Issues

The role of the researcher during the fieldwork period is a complicated one which involves issues of power relations, control, accountability and legitimacy. Accounts of Western field researchers, mostly by anthropologists, have documented many of the ethical and political issues involved in the collection of information in developing countries. The roots of social science research in developing countries are closely entwined with the colonial project and colonial social science. Asad (1973) explores this relationship in the context of the position of Western anthropologists in colonial administrations and universities as an 'encounter' which bore fewer benefits for the subjects of research than for the western social scientists involved.

Bulmer (1993: 12) states that the social survey in developing countries faces problems due to the heterogeneous nature of ethnic, racial, political and linguistic patterns in developing countries and that "the much greater variation (than in the West) within developing countries poses a sharp challenge for the survey researcher which it is difficult to meet." While the caution with which Bulmer gives to the survey researcher in applying survey methods in developing countries is valid in advising against generalised applied frameworks, such a comment also makes an undue assumption that the researcher's position is that of a Westerner with little knowledge of the 'local' setting in which the research is being conducted. Even more importantly, his description of the West as relatively homogeneous ignores such categories as class, gender, race, political persuasion, regional identity and ethnicity which are not simply exclusive to developing countries. A more contentious statement is presented by Wuelker who comments on the compatibility of research methods used in the West with those applicable in Asia:

> Asian peoples require an entirely different approach from that indicated in dealing with the peoples in countries to which industrialization came earlier. Since the number of illiterates in these parts of the world is still large, relatively few people are able to find things out for themselves and, accordingly, the ability to form their own opinions and express them is confined to comparatively small sections of the population...There is much more need for a thorough knowledge of the economic and social structures of those lands. This is a region where fieldwork and scientific research must go hand in hand (Wuelker 1993: 161).

Such methodological and epistemological positions have been formulated on the basis of overtly Eurocentric opinions and assumptions. Bulmer and Warwick (1993: 21) attempt to challenge these assumptions stating that the age has passed of the 'safari' scholar, typically North American and Western European, undertaking research as expeditions of 'outsiders' gathering data of no positive significance to the host societies. The ethical issues involved in conducting research in developing countries, however, represent a distinct discipline in themselves.

The use of research by development practitioners and policy-makers has created an uneasiness among many social science researchers interested in critical studies of development processes (Mikkelson 1995). This uneasiness is illustrated in the manner in which 'development studies' as a discipline has evolved into a broad area of study able to, on the one hand, undertake theoretically critical research, while on the other to also engage in studies which aim to legitimise existing concepts and practice. The relationship between the utility of development research for policy and the potentials of research as independent investigations of social and political development and change is one which has generated debate within development studies. This has most notably resulted in an evolving development vocabulary which has continued to adopt concepts such as 'capacity building,' 'community participation' and 'action research.' The shift in development research to more 'bottom-up' approaches has been widely accepted by local non-governmental activity as well as by international project planning agencies most commonly associated with 'top-down' methods. However, the results of such approaches cannot be generalised here as the nature of the projects, the actors and the funding implications have shown a diversity of trajectories. This study, though of a small-scale and with no attachments to policy agencies, still relates to some of the above mentioned issues.

Cernea argues that the use of social knowledge in development intervention has traditionally occurred through projects. However, sociological analysis offers the opportunity to identify the relevance and applicability of models for intervention without "obliging the 'developers'... to blindly accept the framework" (1990: 7). Mikkelson (1995), however, comments that "anthropology and other development studies must be seen both as a means of intervention and as a means to understand the nature of change." With regard to this argument, this study has not been constructed as a direct planned intervention but as a means of understanding the social dimensions of housing access. However, all studies, whether direct or indirect, are interventions, particularly where the researcher is from outside of the sphere of the group or area being studied, and similarly field surveys are an interference, given the imposition of time and space required for interviews (Mikkelson 1995).

The study's aim was to examine the dynamics of access to housing among Amritsar's low-income population and not to intervene in the capacity of a development agency. While many respondents in the survey initially expressed their expectations that their interviews would result in some form of action, they were told that the information given in their interviews was only to be used for research being done by us at the local university. As a result of the activity of various public agency and political party workers in upgrading private self-help areas and in evicting *jhuggi* settlements, there was a high level of awareness of the possibilities of benefit from talking to 'outsiders.' The limitations of the research can be highlighted within this context.

My experience in Amritsar as a Punjabi/Sikh North American-born researcher from a British university highlights the complex nature of the role of the researcher. I had gone to Amritsar for the purpose of collecting data and information for this thesis. As a person of Punjabi descent, language and ethnicity did not present immediate obstacles to the research, though in the in-depth interviews I was assisted by the research assistants who were more linguistically proficient than I was. Other dynamics of power as a researcher were governed more significantly by class, caste, education and gender (Shahrashoub 1992; Mascarenhas-Keyes 1987). The perception of the researcher by those who are researched effects the way the interaction is defined (Burgess 1984). The ethical dilemma of the privileged researcher interviewing poor, low-caste households was recognised during the fieldwork as a limitation to the study. The main limitations were that my perceived position of power and influence prompted responses which respondents believed would result in direct policy actions to better their conditions. Similarly, my inability as an outsider and researcher during the limited period of time to partake in inducing change in the poor settlements of Amritsar was both a limitation and shortfall of the study. The differences between the outcomes that residents desired from the interviews and what I, as a researcher, hoped to accomplish expose contradictions and inherent limitations to the uses of such a study. Nonetheless, the findings of the study will, it is hoped, have a role to play in furthering the understanding of the social dimensions of shelter policy.

---

[1] The Green Card scheme distributes food and household supplies at subsidised prices to poor households which qualify.
[2] See Appendix 1.
[3] Tipple et al (1994) examine the tendency in housing studies to adopt English language and Western European concepts which, they argue, causes discrepancies between local interpretations and technical, policy orientations. They use examples from two West African countries, Ghana and Nigeria, to illustrate how alternative terms can be defined by researchers which suit the local setting.
[4] Turner preferred to use vernacular terms which focused on the process of settlement such as the *barriadas* in Lima (1972). Schoorl, van der Linden and Yap

(eds.) (1983) and Hasan (1987) use the terms *bastis* and *katchi abadis* in reference to settlements in Pakistan while Paul Baross (1984) uses *kampongs* to describe the settlements in Indonesia.

[5] J. van der Harst uses the same categories in his discussion of investment costs in the slums of Karachi. See 'Financing Housing in the Slums of Karachi' in J.W. Schoorl et al (1983) (eds.) *Between Basti Dwellers and Bureaucrats: Lessons in Settlement Upgrading in Karachi. Jhuggies* are tent, hut structures made of cane, plastic, tins, metal or other scrap materials. *Kaccha* houses are made of mud or clay. *Semi-pacca* houses are strong structures with a solid base with impermanent roof and/or floor material. *Pacca* houses are strong structures with lasting, firm roof and floor materials.

[6] While evidence of other Indian cities shows that the urban poor find housing in squatter settlements, in Calcutta formal rental sector housing within legal and tax regulations is a common type of low-income housing.

5 The Research Context

Introduction

A number of factors have contributed to the development of the low-income housing system in Amritsar such as government policies towards low-income housing and the spatial and social patterns of residential location. As illustrated in Chapter Three, the city of Amritsar has undergone a history of development and decline and this is further reflected in the evolution of an intricate system of shelter provision serving the low-income population. From settlement on free or vacant land by migrants, allotments to new housing stock created by conventional schemes to the improvement of already existing settlements, there have been few choices available to the poor. Those shelter opportunities that have been available have been to a great extent socially determined and differentiated, as the study here argues.

Policies at the national and state levels have shown parallel trends of devolution and withdrawal from housing provision. The state's role as a provider and initiator has been progressively 'down-sized' to that of a supporter of individual household efforts. This directly corresponds with the self-help ethos addressed in the first two chapters which has seen the proliferation of similar trends in many other Third World contexts. How this has impacted upon the local low-income housing system in Amritsar will be examined in this chapter through an identification of the different housing forms in the city.

National Housing Policies in India

Largely due to the post-colonial history of the Nehruvian socialist development model, conventional housing policies in India have, until recently, been the primary housing development policy. The First Five-Year Plan[1] in 1951 commenced the provision of housing for low-income groups with conventional projects through flats and tenements constructed specifically to contribute to the supply of housing (Mehta 1988). However, conventional schemes in India, as in other developing countries, diversified through a number of ventures such as housing for industrial workers in 1952, housing for low-income groups (LIG) in 1954 and housing for plantation workers in 1956 (Ansari 1995). Over the next few decades a number of schemes were launched such as slum clearance and improvement, loans for the construction of units, subsidised housing for industrial workers and social housing for low and middle-income groups. Public conventional housing schemes have until recently been undertaken by the state in India without any involvement by private sector interests.

In India the reservations about conventional policies were promptly accompanied by land controls used as a means for diverting concentrations of land for social ends. The Urban Land Ceiling and Regulation Act (ULCRA) was introduced in 1976 as a means of countering the concentration of land holdings in urban areas. It has since, albeit in a limited capacity, been implemented by local development authorities throughout many urban areas in India. Through the act, private ownership of land is restricted and acquired excess land has been developed and allocated towards public schemes such as housing. However, loopholes in the execution of the act have had several "unintended results"(Raj 1990: 269). First, it has shown to have had reverse effects upon the identification of land where land has been exempted (75 percent of cases) rather than acquired. Another obstacle has been that only a small proportion of successfully acquired land could be reallocated for residential purposes. The lengthy bureaucratic procedures and litigation that have followed acquisition cases has caused many plots of land to be frozen from further development thus resulting in overall increasing land market prices.

The case of Delhi is a perhaps unique one of how large-scale public land acquisition has been carried out over several decades as a measure to 'socialise' urban land through a scheme of direct state intervention and regulation of planning, land-use and development (Mohan 1992). In the 1950s pricing controls were adopted to ensure adequate supply of land and to check speculation of land values for the benefit of the poor. The lowering of land prices through such mechanisms did make land available at lower prices, though the speculative forces were not eliminated and the demands for housing by low and middle income groups was not met (Mitra 1990).[2] As a result the following period of housing policy marked a dramatic shift, according to Harms' continuum, towards state-initiated policies.[3]

The Sixth Five-Year Plan (1980-85) emphasised the importance of affordability in public housing schemes. The plan document reviewed social housing schemes in terms of their accessibility to the specified low-income categories. It was found that households in higher income categories were occupying houses built for lower category households due to the inaffordability of instalments required (Bhattacharya 1990). The Sixth Five-Year Plan therefore encouraged self-help and the provision of infrastructure as a means for stimulating the private sector's involvement in housing development. The plan emphasised the need for cost reduction and the lowering of standards in order to cut expenditure on costly materials and procedures. The housing issues that were identified during the Seventh Five-Year Plan (1985-90) carried on with this approach by affirming minimum standards and the passing on of responsibilities to the private and households sectors (Bhattacharya 1990).

A further change in national policy is reflected in a shift from a state-initiated to a state-supported approach. The National Housing Policy of 1988 and Seventh Five-Year Plan (1985-90) both emphasised the importance of access to finance through the expansion of funds from both formal and informal sources. Central and state government budgets would be allocated in priming the housing sector while financial organisations such as the Life Insurance Corporation of India (LIC), Unit Trust of India (UTI) and commercial banks would provide the basis for the distribution of available funds for household financing. A number of specialised institutions such as the Housing and Urban Development Corporation (HUDCO), co-operative finance societies and private sector finance companies such as the Housing Development Finance Corporation (HDFC) were encouraged to plan and mobilise resources for the housing sector. This co-ordination of private and public sector agencies has since dominated the approach towards housing development in India. The Eighth Five-Year Plan has since furthered the private sector's role in housing development through the utilisation of the Improvement Trusts.

State Housing Policies in Punjab

The trends of the national housing policies in India are also reflected in the policy actions of the state of Punjab. The Punjab Housing Development Board (PHDB) was constituted under the Punjab Housing Development Board Act in 1972 for the development and implementation of a number of ventures: housing programs for Low-Income Groups (LIG's) and Economically Weaker Sections (EWS), management of Urban Estates and the regulation of a number of acts under the department of Housing and Urban Development. The Punjab Urban Planning and Development

Authority (PUDA) oversees and manages all activities of the PHDB as an umbrella organisation for nearly all urban development activities in the state. The Urban Land Ceiling and Regulation Act, which is implemented by the Urban Development Wing of the PHDB, only applies to the three largest urban areas in the state: Amritsar, Ludhiana and Jullundur.

In June 1988 the National Housing Policy was accepted by the states as a guideline for future housing development. The recommendations of the National Housing Policy were that the role of public agencies should be kept minimal as facilitators of development rather than as producers of housing. Within this strategy both formal and informal sector actors were invited to take part in the provision of housing in order to accelerate the process of housing production. This was to encourage housing development not only for high income groups, but also for middle and low-income groups. The construction of public housing schemes has traditionally been initiated by the Housing Board. However, after the National Housing Policy in 1988 the state of Punjab's urban development bodies were reorganised to accommodate the integration of public and private partnerships. As a result, all activities of the Housing Board as well as those of all other urban-related activities were brought under the jurisdiction of the PUDA. Thereafter, the state's overall approach towards housing development, in reference to Harms' continuum, moved from state-initiated to state-assisted methods.

This shift towards a state-assisted self-help approach is also evident in the attention drawn to household finance and loans. In Punjab two particular agencies have been established to provide financial provision for assisting poor families in acquiring housing. The Punjab Scheduled Castes Land Development and Finance Corporation (PSCFC) and the Punjab Backward Classes Land Development and Finance Corporation (BACKFINCO) have both been appointed as integral elements of the state's poverty alleviation programmes in distributing soft loans and bank loans to scheduled caste and poor households (Government of Punjab 1994-95). These finance organisations were set up as intermediaries between banks and the poor to facilitate the access of the poor to financial resources. However, the disbursement rate of funds has not exhibited a significant number of beneficiaries to such programmes.[4]

The implications of such collaborations between the public and private sectors have shown a general change in policy-making decisions in the low-income housing sector. The division of responsibilities between the formal and informal sectors follows a broader decision by policy-makers that the task of decreasing the housing gap was too mountainous for public sector agencies to tackle alone. Therefore, the direct involvement of the public sector in housing construction has been made minimal while self-help and private sector efforts are to be encouraged. The role of public sector agencies thus transformed from housing constructors to facilitators of housing development. Because of the inherent dangers of exclusion of the economically disadvantaged, one of the public sector's remaining tasks is to check on the servicing of Low-Income Groups (LIG) and Economically Weaker Sections (EWS).

As part of India's five-year plans the state governments were provided with financial assistance from the central government for the implementation of slum clearance programmes, something that had been passed by the Punjab government in 1961. The emphasis was on the removal of slums and the construction of new tenements for relocation, though only a few hundred new tenements were actually constructed and only a small number of slum dwellers were given alternative accommodation. With the National Housing Policy and Report of the National Commission on Urbanisation in 1988, the slum eradication policies of the 1960s and 1970s were officially replaced by a more holistic 'environmental improvement' approach. The scheme of Environmental Improvement of Slums (EIUS) had been introduced in the Fifth Five-Year Plan (1974-79) as a means for both dealing with poverty and developing tools for self-reliance among weaker sections of society (Ansari 1995). The EIUS program corresponded with the growing international trends which were lobbying for the acknowledgement that slums provide a vast proportion of the housing stock. The Governments of India and Punjab have since adopted the Urban Poverty Alleviation Schemes (UPA) which have stemmed from the environmental improvement approach. Within these schemes, upgrading through incremental infrastructure development is one of the most recent policies. The Urban Poverty Alleviation Schemes, namely the Urban Basic Services Program (UBSP), are centrally sponsored development schemes which specifically target low-income localities by providing infrastructure such as water, sewerage, street lights, latrines and storm water drains.

Few studies have been done on the impacts of slum upgrading in Punjab. However, Sandhu's (1989) study of the effects of the Slum Improvement Programme in Ludhiana offers insights into some of the results of the upgrading process. In the study it was found that there were drawbacks to the implementation of the scheme. Lack of enthusiasm by the Municipal Corporation, inefficient co-ordination among the agencies involved, shortage of funds and sub-standard maintenance of improved areas were a few of the failures that were identified, though a general improvement of physical infrastructure was a noticeable benefit of the programme. However, common to the wider experience of upgrading programs, only a negligible number of residents actually benefited from the improvement schemes (Sandhu and Sandhu 1988).

A more recent, and less documented, policy initiative is the introduction of public agencies as developers of commercial housing projects. With the emphasis on cost-efficiency in housing development in previous years, the state apparatus has begun to use its monitoring abilities to stake its position in housing and land development. Private and informal speculation of land prices has been a long-term obstacle to the accession of low-cost land for housing the poor in urban areas (Angel et al 1983; Raj 1990; Baross and van der Linden 1990). Improvement Trusts have been established as the designated government-related bodies which announce the allotment of vacant or built-up plots to the public. While these schemes are not limited to low-income groups, the principle of cross-subsidisation is upheld such that those commercial sights which are sold at high prices to commercial interests or high-income groups will subsidise the costs of smaller plots targeted at low-income allotees (Rao 1991). Whether this principle, as was attempted in public housing schemes, ensures accessibility to the poor will provide another dimension to the research on housing development. This new form of housing development, while slightly detracting from the preceding incremental improvement policies, exhibits the state's increasing role in commercial ventures and speculation.

The diversification of central and state government activities in housing development in Punjab is exemplified in the types of low-income housing settlements in Amritsar. A number of public agencies have been involved in conventional, state-assisted and state-initiated schemes for low-income groups: the Municipal Corporation, Punjab Housing and Development Board and the Amritsar Improvement Trust. The extent to which each scheme has utilised public and private resources varies and demonstrates the increasingly varied types of responses to housing for low-income groups as well as the number of different agencies which have become active in low-income housing development.

Low-income Housing in Amritsar

The national housing policies in India have shown a dramatic change in approach over the past few decades from conventional projects to state-supported self-help schemes. Similarly, the housing policies in the state of Punjab have also followed the same line of increased participation of the private sector and individual self-help in state-supported programs. In Amritsar the recent history of housing policy towards the poor exhibits a number of parallel trends which have shaped both the formal and informal elements of low-income housing in the city. In this section, the experiences of housing policy in Amritsar will be first examined followed by an analysis of the low-income housing system in operation in the city.

In 1975 the Town and Country Planning Department, on the basis of the Punjab Slum Areas (Improvement and Clearance) Act, notified 19 'slums' with an estimation of slum dwellers at 32,632 in Amritsar which increased to 63 in 1991 (Figure 5.1).

Figure 5.1 Notified Slums of Amritsar

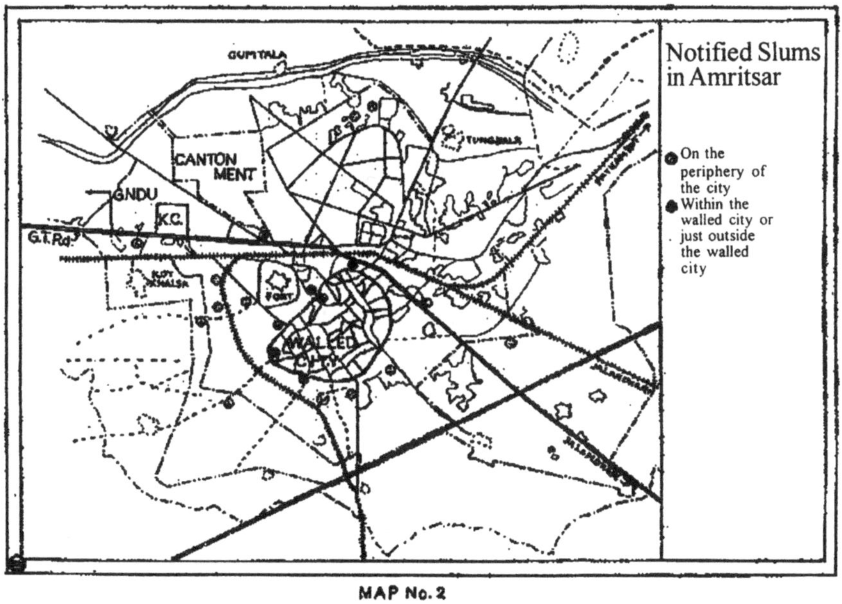

MAP No. 2

Source: Sandhu (1989) pp. 28-29.

The Amritsar Municipal Committee was elevated to the status of Municipal Corporation during this time which extended the municipal boundaries of the city affecting these figures (Sandhu 1989). These figures, of course, only include those settlements which have been officially declared - the actual figures here estimated to be much higher. The definition of 'slum' as applied here is defined as an area which lacks one

91

basic amenity (Amritsar Municipal Corporation 1993). Identification of 'slum areas' in Amritsar has more implications than simply the inclusion of notified areas on official lists. Recognition of areas also relates to the intentions of official bodies towards these areas. Once an area is notified it is assumed to be legalised, or at least not to be evicted in the near future.

The housing system which services the poor has a number of attributes which extend beyond the physical traits of localities. Patterns of land access and ownership reveal the structural and legalistic obstacles faced by poor communities. The manner in which land is settled upon and developed has a direct impact upon the paths to housing that are available to poor households. As just mentioned, the low-income urban population has proportionately increased from 19 settlements of a population of 32,632 in 1976 to 63 unauthorised settlements in 1991 (Municipal Corporation 1993). However, this survey done in 1995 estimates a higher figure, on the basis of income and housing condition and services, which is estimated to be close to fifty percent of the total population figures. The low-income housing system in Amritsar is made up of four sub-systems: the marginal system, the informal system, the pre-colonial system and government *ad hoc* projects. The characteristics of each system have been determined by a number of interrelated factors such as process of settlement, construction and patterns of redistribution of houses.

The marginal system has been defined in Marcussen's (1990: 43) study of Jakarta as "an extra-economic system structured by social relationship with regard to settling, building and redistribution...Members of the colonies are low-income households and the percentage of recently arrived migrants or temporary migrants is higher than city averages." As in Jakarta and other Third World cities, the marginal system in Amritsar is articulated through squatter settlements (*jhuggi* settlements) which are marginal not only in economic aspects, but also spatially and legally. *Jhuggi* settlements in Amritsar are distinguishable as they exist on unserviced tracts of land either on the peripheries of the city or on land which is undesirable to other groups. *Jhuggi* settlements have generally developed through squatting upon public land, and the social structure of the settlements show strong kinship links. The make-shift structures are the only material possession that *jhuggi* residents have claim to. When a household leaves the settlement, they generally cannot sell their home since the land is illegally occupied, but the materials with which the house was constructed are an economic asset to the household.

The informal system of housing in Amritsar has, until recently, existed purely outside of legal frameworks. With the shift in policy approach away from conventional methods, the informal system has become increasingly integrated into the government's planning structure. As briefly discussed in Chapter Two, this recent development corresponds with the wider emergence of unregulated markets in many Third World cities. The Municipal Corporation, in its acknowledgement of the considerable contribution of the informal system in providing housing and in its unwillingness to undertake more conventional schemes, has begun to give assistance to self-help settlements which either lack legal tenure or service provision. The upgrading of illegal settlements has been a slow process in reaching many areas, though often upgrading has occurred where legal tenure has not been officially declared.

The pre-colonial system of the old walled city is not specifically being investigated in this study, although its own history of development is central to the existence of low-income settlements around the walled city. The walled city represents the oldest housing stock in Amritsar which was developed and consolidated from the sixteenth to nineteenth centuries before British rule. Unofficial estimates of the total population of the walled city are around 250,000. Due to the expansion of the city, the centrality of the walled city to the overall low-income housing system has declined over the past three decades. The multi-storey houses are today occupied by families who have traditionally lived in their ancestral homes, partition refugees as well as low-income households who cannot afford to either buy or rent houses elsewhere. Incremental development of the main areas is visible throughout the walled city, though this is mainly in commercially dynamic areas within the walled city.

As discussed earlier in this chapter, there are a number of public actors in the realm of housing development in Amritsar which have contributed to the government *ad hoc* system of low-income housing. The Municipal Corporation oversees the legal and maintenance aspects of public housing while the PHDB and Amritsar Improvement Trust are engaged in new developments. Previously, the Municipal Corporation had constructed tenements for its employees. The experience of these tenements showed that they were neither of adequate quality nor were allocated to the appropriate targeted groups. Since then, the responsibility of executing public schemes has been taken by the PHDB and Amritsar Improvement Trust. The nature of these projects is that they are developed for specifically targeted income groups. High, middle and low-income groups are all included in these schemes which are financially planned according to cross-subsidisation between income groups. This system of housing operates, in theory, through tenders and official applications in order for applicants to be assessed on their eligibility for the schemes.

Land System

With the growth of cities and the increasing commercial and residential demand for land, land has become more than just a place for settlement; it has become an economic asset. However, where customary land systems are still in existence commercial effects may be curtailed, though even this type of land system is becoming scarce. The commodification of land means that, in simple terms, the builder of a house must pay for the land at a price set by the local land market. Non-commercial articulations refer to instances where the potential house builder does not have to pay for the use of the land, or in the case of customary land practices, the person would give payment as a 'voluntary gift' rather than in monetary form (Baross 1983).

Baross (1983) describes the ways in which land is accessed in urban areas for building houses in Jakarta, Indonesia. He uses the concept of 'social articulation' to illustrate the various forms of land supply: non-commercial, commercial, and administrative. In Amritsar all three forms of articulation are evident in the overall land market. However, the commercial forms have begun to push the poor out as rents and sale values have become unaffordable to them. The non-commercial sector until now has served to absorb these people by offering them land for housing near to the walled city and free of cost. From these settlements those with enough savings to afford flats in government subsidised schemes have been offered a route to serviced housing through formal bureaucratic procedures.

One of the long-term effects of colonialism has been the inheritance of a dual land system, which in Amritsar has resulted in decentralisation of monitoring land use. All lands can be categorised as either 'registry lands' or 'public lands.' The present formal land system was created by the colonial administration and operates under the Municipal Corporation's jurisdiction and distribution of registries (or titles) denoting private or collective ownership. Public lands are those which are held by public agencies for which no registry is held, though the official status is recorded by the Municipal Corporation. Until 1976, no restrictions had been placed upon the holding of land, and the patterns of land-use went virtually unchecked. Land owned by public agencies became either utilised for local government activities or left undeveloped while private land has been more actively developed for residential and

94

commercial ventures. The land ownership around the walled city in zone 2, holding the highest concentration of low-income settlements in the city, is distributed among a number of public agencies.

Figure 5.2 Land Ownership

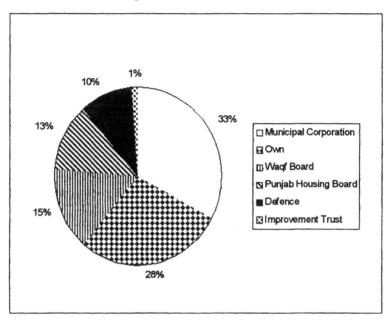

The Municipal Corporation is the local government body which oversees all legal and state frameworks at the city-wide level. It is the largest public land owner in the sample with 33 percent ownership outside of the walled city. Private owners are the second largest owners, with 28 percent identifying themselves as owning the land on which they are living. The administrative element of ownership of land in Amritsar is done through the registration of land with the Municipal Corporation and in the granting of titles. This system was introduced by the British in the recording of all property-related transactions. Customary land systems, however, are still operating as a parallel system through the inheritance of ancestral land and the disbursement of land through the Waqf Board.[5] The Waqf Board, was an early non-British formal institution which was created in order to manage the land holdings which were to be distributed among the beneficiaries. It owns land which, for purposes of comparison with other cities, resembles the customary land systems of many Third World cities where pre-colonial land patterns are still in operation. The more integrated nature of Waqf

Board transactions with the formal local government institutions since independence reflects the attempts by the government to abolish the dual system of formal and customary land use patterns. Instalment arrangements and, in some cases, the sale of land to residents living on Waqf Board land exhibits, as in other Indian cities, the new hybrid form of land use which has emerged from post-independence urban development (Bradnock 1984; Zetter 1984).

Spatial Pattern of Low-income Housing

The typology of *jhuggi*, private self-help and state-assisted settlements developed in this chapter provides a framework for describing the basic low-income housing system in operation in Amritsar. Each type of housing has its own history of early settlers, consolidation and socio-economic characteristics which allow for a qualitative assessment of the dynamics of social access in Amritsar's low-income settlements.

Fifteen settlements have been surveyed in this study, all of which are located around the walled city[6] (Figure 5.3). In the following empirical chapters, reference to individual settlements will be made as illustrations of particular aspects of housing access.[7]

Figure 5.3 Survey Settlements

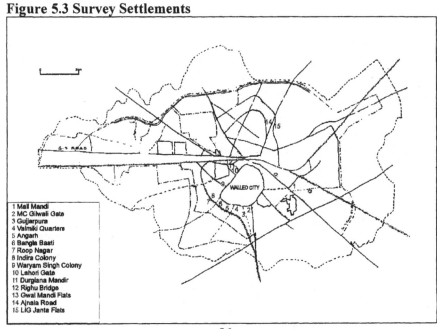

1 Mall Mandi
2 MC Gilwali Gate
3 Gujjarpura
4 Valmiki Quarters
5 Angarh
6 Bangla Basti
7 Roop Nagar
8 Indira Colony
9 Waryam Singh Colony
10 Lahori Gate
11 Durgiana Mandir
12 Righu Bridge
13 Gwal Mandi Flats
14 Ajnala Road
15 LIG Janta Flats

On the western side of the walled city as well as on the peripheries of the city, squatter settlements have had a history of settlement, eviction by the authorities and rebuilding of *jhuggies* by residents. These squatter settlements have generally been home to newly arrived economic migrants as a 'first stop.' On the western side of the walled city almost all low-income housing is built on disputed public land. These are predominantly owner-built houses where registries have only recently been granted. Signs of commercialisation in these areas are evident with reselling of homes, a trend which may increase as security of tenure is granted. There are a small number of renters in these settlements.

One of the survey settlements, Mall Mandi, is situated on what had previously been agricultural land but which has increased in value due to its location just off the Grand Trunk road and in the gradual commercial expansion on the eastern side of the city. After the storming of the Golden Temple in 1984, due to pressure from businesses which had been displaced by Operation Bluestar, the central government began the Galiara Scheme which was meant to deliver compensation to displaced businesses and also to 'clean up' the surrounding areas of the walled city, particularly the underdeveloped eastern side. This area had historically been home to several squatter settlements due to its previously low commercial value. The Galiara Scheme, while increasing the demand for land in this area, also resulted in the eviction of hundreds of households through the relocation of businesses from inside the walled city. Many of these households were those which had been threatened during the Emergency in 1975-77 when the then Prime Minister Indira Gandhi announced that all unauthorised occupation of vacant land would result in evictions. This affected settlements on both publicly-owned and privately-owned land where land owners were given public support to evict unwanted squatters.

The low-income areas on the southern side of the walled city comprise of a combination of Municipal Corporation housing and owner-built, owner-occupied partition settlements primarily built on Waqf Board land. These settlements are static in terms of their development, though they face no real threats of eviction. The upgrading and legalisation of many informal settlements in the area has resulted in a sense of security among residents and in the further consolidation of such settlements. In present day Amritsar, low-income areas are consolidating and slowly expanding along the walled city, particularly around the north, west and southern sides of the outer ring road surrounding the walled city. As property and land values continue to rise as the city enters a new era of political stability, it is likely that commercial interests in the area will try to push these households out of their homes. The effects of commercial pressures will no doubt effect all types of housing, from *jhuggi* settlements which will be most vulnerable to eviction to state-assisted housing where the poorest may be excluded from occupancy. For the present time, however, poor settlements will continue to live on the margins of the walled city.

---

[1] India's development strategy post-independence has been executed through five-yearly economic strategic plans with the first starting in 1951.

[2] One of the most profound effects of such measures has been in the dramatically high proportion of people in Delhi living in informal settlements, including slum and squatter settle ments (Gupta 1985; Mitra 1990).

[3] See Chapter Two for a more in-depth look at Harms' continuum.

[4] In 1993-94 the PSCFC disbursed Rs 21.74 crores to 1994 Scheduled Caste beneficiaries while in 1992-93 it has given Rs. 27.68 to 25122 families.

[5] Since independence the Waqf Board has become a formal institution acting within the formal land system.

[6] Several public housing schemes in the state-assisted category exist slightly outside of zone 2 but nonetheless have been included within the study due to the significance of state activities in low-income housing provision.

[7] See Appendix 2.

6   Socio-Economic Profile of the Sample

Introduction

The structurally marginal position of the poor, predominantly low-caste, communities in Amritsar cannot be addressed without considering the social, economic and political obstacles which they are confronted with. A number of factors will be identified in this chapter regarding the structural position of the poor. The relationship between caste and religious identities with employment and migratory processes, though changing, continues to have significance to the material well-being of the poor. For instance, the experience of partition refugees, even to today, has marked them as a distinctly separate community from other more recent migrants from other parts of India. On another level, the conflation and overlapping of various social identities, as will be addressed in this chapter, have revealed a highly differentiated social structure within low-income communities in Amritsar. This, in turn, has had repercussions on occupational opportunities, income security and the scope for upward mobility within these communities.

Caste

Caste is a form of social organisation which, while having experienced a number of changes due to modernisation and urbanisation, is still a dominant part of social, economic and political life today not only in Amritsar but in the rest of India (Bradnock 1984). According to the 1991 Census of India the total national population was counted as 838.58 million. of which 138.22 million (16.48 percent) were scheduled castes.[1] Caste groupings and categories are positioned within the Hindu *varna* system according to beliefs of ritual purity and social order (Wiebe 1975). Brahmins are positioned at the top of the hierarchy, traditionally associated with religious and learned professions while scheduled castes have been discriminated against through constructions of purity and servility which have placed them at the bottom of the hierarchy.[2] Caste as a social category is defined as homogenous groups which practice endogamy and common occupations. There are regional differences which show variations in caste hierarchies relating to interaction as well as political and economic variables (Wiebe 1975; Mayer 1980). Despite changes to caste in modern India where a certain amount of professional mobility has undermined the traditional notions of caste as fixed categories, it continues to have significance to social organisation and hence is an important social grouping by which social access in this study can be measured. Particularly since a vast majority of low-income residents belong to lower castes, an examination of the representation of caste is useful.

The figures on the state of Punjab show that the distribution of scheduled castes within the total population are much higher than the national average.

Table 6.1 Distribution of Scheduled Caste Population in the Five Highest Ranking States

State	1981	1991
Punjab	26.87	28.31
Himachal Pradesh	24.62	25.34
West Bengal	21.99	23.62
Uttar Pradesh	21.16	21.04
Haryana	19.07	19.75

Source: Census of India (1991).

Punjab ranks as the highest state in terms of percentage distribution of scheduled caste population in India. This can partially be accounted for by the heavy influx of migrant labour, a large number of whom belong to scheduled castes, from Bihar and Uttar Pradesh, over the past two decades (Singh 1997). These figures cannot be separated from the connection between economic position and caste background. While the significance of caste in different regions within South Asia may vary, an assumption is made here that economic position is considerably affected by the rigidities of caste hierarchies on the lower castes and the obstacles that they present to economic upward mobility.

The sample survey in this study specifically targeted low-income housing settlements, and therefore the high representation of scheduled castes in the survey is derivative of the sample selection (Figure 6.1). The largest caste group represented in the survey of Amritsar are Majbi Sikhs who represent the effects of partition upon the demographic make-up of the city, as discussed in Chapter Three. Scheduled castes and Valmikis also form a considerable proportion of the low-income settlements around the walled city.

Figure 6.1 Caste Make-up of Sample

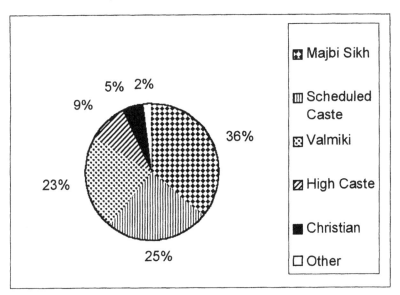

The category scheduled caste is not exclusive of other categories listed in the caste make-up of the sample. In the 1991 Census of India the total number of scheduled castes or groups of castes was estimated to be 1091 out of which 37 are notified as being in Punjab. The category Scheduled Caste in the census is a more generalised term than as it is applied in this study. Scheduled castes are listed in the census on the basis of their relationship to the Hindu caste (*varna*) system. In contrast, I distinguish scheduled castes from other lower castes. For example, Majbi Sikh, Valmiki, and Christian are identities which are closely tied with the Hindu *varna* system. A vast majority of Christians and Majbi Sikhs were originally Hindu scheduled caste groups who converted to Christianity and Sikhism in order to, as they believed, rid themselves of caste oppression.[3] However, the extent to which emancipation from caste discrimination has occurred through conversion to Sikhism and Christianity in the Punjab context is not direct, given the identification of such communities in Amritsar's low-income settlements. The Valmiki community here has been defined as both a caste and a

101

religious group.[4] While this community's position within the Hindu *varna* system, as with Majbi Sikhs and Christians, may fall within the scheduled castes, they also constitute separate caste identities as they form distinct communities in the Punjab context.[5]

Religion

From the previous analysis of caste, it is clear that caste and religion merge as overlapping identities in India. This is particularly the case in Punjab where the influences of Islam, Christianity, Buddhism and Sikhism have resulted in the mixing of identities. Punjab was at the centre of the *bhakti* movement of the fourteenth and fifteenth centuries at which time religious tolerance and coexistence were being challenged by a resistant and disparate Hindu system and a powerful Mughal empire. The *bhakti* movement signified the 'meeting' of Hinduism and Islam, out of which was born Sikhism, a religion indigenous to the region of Punjab. Colonial rule resulted in the communalising of identities (Oberoi 1994), divisions which became further deepened by the partition of the region in 1947. The partition of Punjab saw the division of Hindu, Sikh and Muslim communities according to the agendas of the emerging post-colonial states. The representation of religious groups in Punjab show the effects of partition, particularly with reference to the low proportion of Muslims (Table 6.2).[6]

Table 6.2   Representation of Religion in Punjab

Religion	Population	Percentage
Sikhs	12,316,480	61.0
Hindus	7,404,870	36.67
Christians	222,845	1.10
Muslims	202,550	1.00
Jains	32,590	0.16
Buddhists	960	0.005
Other religions	9,235	0.045
Religion not stated	1,265	0.006
Total	20,190,795	100.0

Source: Census of India (1991).

The religious make-up of the sample in this study is reflective of the above-mentioned history. While castes are, by definition, linked to occupations through 'divine decision', occupation is also religiously associated with caste

102

(Mayer 1980: 56). The religious and caste identities together have formed the basis for community groupings which are significant in present day Punjab. As with the high proportion of Majbi Sikhs in the caste make-up of the sample, the religious background of respondents similarly shows the major group to be Sikhs.

Figure 6.2 Religion of Respondents

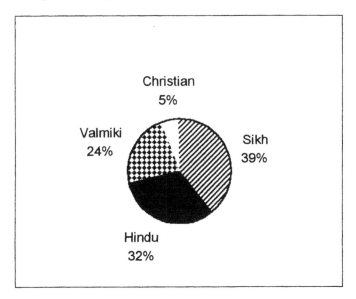

A majority of Hindu respondents lived either in or around the walled city prior to moving to their present houses. As is particularly shown in the case of Lahori Gate, a majority of Hindus living in this settlement moved into their homes at the time of partition before Hindu and Sikh refugees from west Punjab could occupy them.

The high representation of Majbi Sikhs in the sample is countered by the absence of Muslims. There is a small proportion of Muslims living in Amritsar, who are not visible in the low-income settlements around the walled city. Most of the Muslim community in Amritsar moved west to Pakistan as a result of partition in 1947 which explains for the relative invisibility of Muslims in the city. From being the majority religious group prior to 1947 to becoming a present day sparsely numbered minority, the Muslim community in Amritsar today only exists as pockets of artisan groups within the walled city who predominantly work in the embroidery and textile trades.[7] However, none of these communities reside within the survey sample settlements and therefore are not represented in this study.

Migration

While Amritsar is not a city which has experienced rapid urbanisation in its recent history due to its historical decline in comparison to other South Asian cities such as Ludhiana (Oberai and Singh 1983), Karachi (Nientied 1982; van der Linden and Selier 1991) and Delhi (Pugh 1990), migration nevertheless plays a significant role in the social make-up of the low-income settlements taken in the sample.[8] The proportion of migrants is 41 percent compared to 59 percent non-migrants in the sample. This is significant as Amritsar is not associated with high levels of urbanisation as is shown in Chapter Three in the census data available on Amritsar.

Table 6.3 Migrants and Non-Migrants

	Jhuggies	Private Self-Help	State-Assisted	Total
Migrants	100.0	33.0	48.0	41.1
Non-Migrants	--	67.0	52.0	58.9
Percentage in Sample	6.5	67.6	25.8	100.0

A number of studies on different urban contexts have widened the understanding of migration. Much attention has been drawn to the diversity of migratory patterns that have formed in different Third World cities (Todaro 1994). Permanent and temporary forms of migration have shown to be consequences of rather differing rural and urban circumstances. Whereas migration was once thought to be the result of households moving from their place of origin to another place where they sought to develop their future, migration has now come to include a range of different types of movement. Lewis (1954) developed one of the earlier models which considered migration as an equalising mechanism by which sectors experiencing labour-surplus could fill labour-deficit sectors. This model is grounded in a dual economy approach which separately distinguishes the subsistence, agricultural sector associated with unemployment from the modern industrial sector associated with full employment (Oberai and Singh 1983).

Todaro (1969) questioned the unilinearity of previous studies which had failed to recognise the multitude of variables involved in the migration process. He argued that migrants choose between the expected gains of migration over the actual earnings in their rural settings. Such expected gains of migration are measured by the difference in real incomes between rural and urban employment opportunities and the likelihood of obtaining a job (Todaro 1976). The models of the 1960s have been attributed to the colonial legacy in the widening rural-urban divide which has hastened migration towards cities, as in Todaro's model, and generated an 'urban bias' which inherently exploits and excludes the urban poor (Todaro 1994; Lipton 1977; Amis 1990). The limitations of such income differential models are that they fail to recognise the diversity of skills and attitudes of migrants. They also assume that all migrants have access to adequate information regarding their decision to move and the probability of finding employment in the urban sector (Oberai and Singh 1983).

Other more recent empirical studies have made a number of interesting findings. Selier (1991) found in his study of Karachi's *katchi abadis* that there were four types of migratory status (which he labels 'mobility'): commuting, circular migration, working-life migration and lifetime migration. In his analysis he quantifies the term 'mobility' in reference to respondents' relationship with their village, residence and work place. Particularly in African countries, circular migration has been most notably detected where a circular lifestyle has been forced upon communities through the introduction of cash economies and organised labour recruitment (Amis 1990).

Oberai and Singh (1983) focus upon the city of Ludhiana and the impacts of urbanisation and migration upon social and economic change. The rapid growth of agriculture in Punjab has led to what Oberai and Singh refer to as a 'rural-urban interaction'. The specific types of migration identified in the sample will be discussed in the following section on reasons for migration. However, the levels of in-migration and urbanisation experienced by Ludhiana are not comparable with Amritsar on a direct level. As discussed in Chapter Three, Ludhiana has ascended over the past two decades as the commercial and industrial centre of Punjab, overtaking Amritsar as the region's dynamic urban centre. Therefore, the case of Ludhiana offers only limited comparative analysis with regard to migration in Amritsar.

Despite its meagre migration experience in comparison with other cities, even a medium-sized city like Amritsar shows a considerable representation of migrants within its low-income settlements. Seven different categories have been constructed around the place to where respondents claimed as their region of origin. Table 6.4 shows the distribution of the regional background of respondent households.

Table 6.4   Place of Origin

Place of Origin	Jhuggi	Private Self-Help	State-Assisted	Total Percentage of Sample
Amritsar rural	5.6	8.1	16.9	59.0
Amritsar	--	66.7	53.5	13.0
West Punjab	16.7	17.7	1.4	10.0
East Punjab	5.6	5.4	9.9	6.5
Border Village	--	0.5	5.6	5.8
Eastern Regions of India	72.2	0.5	7.0	2.0
Other	--	1.1	5.6	1.8
Percentage of Sample	6.5	67.6	25.8	100.0

Those people who stated *Amritsar* as their original home consisted of 59 percent of the sample. The second largest category of 13.5 percent and the largest category of migrants consists of those people who had come from *West Punjab*, namely from the part of Punjab which became divided by the partition in 1947 and which falls within present day Pakistan. Within this category Sialkot, Gujjerkhan and Lahore are the three most common places from where partition migrants have come.[9] Those people who have come from rural villages near and around Amritsar (*Amritsar rural*) make up 10.2 percent of the sample. A cross examination of these migrants with reasons why they have migrated to Amritsar shows that political violence due to the anti-state insurgency and police backlash is a main factor. The areas around Amritsar of Tarn Taran, Gurdaspur district and Ajnala are the main sources of this type of migration.

East Punjab is a category denoting those people who have come from other parts of Indian Punjab. 6.5 percent of respondents claimed to be from other parts of East Punjab which are not near Amritsar. These migrants are primarily from rural areas in Hoshiarpur, Jullundur and Kapurthala districts. The relatively low ratio of this group of migrants can best be explained for by the dominant role that Ludhiana has occupied as the region's economic capital. The rapid growth of manufacturing industries in Ludhiana has seen the city of Ludhiana's population grow by six times between 1941-1981 with 30.15 percent of this growth being accounted for by in-migration (Sandhu 1989; Christopher 1986). Ludhiana has thus had a magnetic effect in attracting labour from other parts of India, particularly from eastern parts of India. The category *eastern regions of India* marks

106

migrants from states in the eastern regions of India, namely Bihar, Uttar Pradesh, Assam, West Bengal and Orissa. These people have migrated to Punjab for the vastly higher economic opportunities and, only because of previously established relationships, have chosen Amritsar over Ludhiana for settlement.

Respondents who fall under the category of *border village* have come to Amritsar for one of two reasons. During the Indo-Pakistan war between 1965-67 those people who owned land along the India-Pakistan border were forcibly displaced by the army and were given compensation for their land. Around 1965-66 heavy rains caused massive flooding which destroyed crops and homes for which the government also gave compensation for resettlement in Amritsar. Waryam Singh colony was built especially for the purpose of housing those people displaced by the floods. While the ratio of 2.2 percent is relatively low to other marked categories of place of origin, these migrants reveal an important element of the Municipal Corporation's past actions around housing provision. The category of *other* contains areas in north India from where people have migrated, though on a small scale of 1.82 percent: Kashmir, Haryana, Himachal Pradesh and Delhi. The 1984 anti-Sikh riots and the volatile political climate in Kashmir in recent years are two reasons given for migration to Amritsar within this group.

As can be seen in the profile of place of origin within the sample, people who are originally from Amritsar make up the majority. Particularly in the case of private self-help settlements, these people are primarily from inside the walled city, many of whom moved as a result of the available vacant homes due to partition. Partition migration and migration from rural areas around Amritsar are also significant indicators of the demographics of low-income settlements in the area. The high concentration of partition migrants in certain private self-help settlements such as Gujjarpura, Angarh and Indira Colony shows that this was the most readily available type of housing for low-income groups at the time of partition.

Table 6.5 Distribution of Partition Migrants

	Jhuggies	Private Self-Help	State-Assisted
Partition Migrants	7.9	89.5	2.6

Migrants from rural areas around Amritsar show the 'pull' effect that urban Amritsar exerts upon its surrounding areas, both due to political instability in the rural areas after the events of 1984 and the perceived economic opportunities that the city offers. Here Todaro's model of perceived gains of migration is evident with regard to the economic migrants. However, the capacity of the city's formal, and less so informal, sectors to adequately fulfil these perceived gains is questionable. The association between the primary reason for migration and other possibly connected variables, such as time of arrival, origin and occupation, may not be grounds for conclusions to be drawn about the role of economic hardship in the place of origin and the prospects of job opportunities in the city (Desai 1995). However, a combination of different types of migration in the sample account for the relatively high ratio of migrants.

Table 6.6 Reasons for Migration

	Jhuggies	Private Self-Help	State-Assisted	Total
Employment	83.3	30.5	42.4	42.7
Partition migrants	16.7	57.6	3.0	34.5
Political Violence	--	10.2	45.5	19.1
Other	--	1.7	9.1	3.6
Percentage of Migrants in Sample	15.9	54.0	30.1	41.1

A large majority of migrants, when asked their reasons for migrating, give the improved prospects of economic standards in the urban economy as the main impetus (Gilbert and Gugler 1982). This concurs with the findings of this study where two predominant types of migration are apparent: economic labour migration and forced political migration. Economic migration comprises 41 percent of the migrant sample. Employment was the response with the highest frequency. While the representation of economic migrants is substantial, it is more a reflection of the declining conditions in the countryside than of the demand for labour in urban Amritsar. The employment structure of Amritsar has not been adequate in absorbing low-income employment seekers as Ludhiana has been. Despite the declining economic record of Amritsar and the lack of availability of secure primary sector income opportunities, the city's largest group is economic migrants, though a majority of these people migrated during the 1970s when Amritsar was yet the state's largest city. Within this group are recurrent migrants who stay in Amritsar for most of the year and return home generally during the rainy season. This resembles the circular migration that van der Linden and Selier (1991) identified in Karachi as an intermediate between lifetime migration and commuting.

While the absorption of incoming labour may not be sufficient in the city and the wages low, the average earnings in comparison to rural areas are most likely higher. Majumdar (1978) in a study of squatter settlements in Delhi found that the average earnings were around two and half times higher that what migrants would have earned in their villages, despite the fact that employment tended to be as casual labourers.[10] Nearly all of the *jhuggi* settlements in the sample are entirely made up of economic migrants primarily from eastern regions in India who have consolidated their communities through friends and family contacts and who have settled in Amritsar even up until recently.

Forced political migration is evident in 1984-affected households and displaced people from the Indo-Pakistan war, together forming 19 percent of the sample. Those people who have come to Amritsar due to political violence are scattered throughout the self-help and state-aided typologies. These people have settled in Amritsar, many not permanently, as a result of the instability caused by violence between anti-state separatists and the police. The areas of Tarn Taran and Gurdaspur are logically the main sources of this type of migration since these regions were the most severely affected after the storming of the Golden Temple in 1984 and the political backlash which followed.

Partition is the second highest response with 34.5 percent of migrants identifying themselves as partition migrants. A limitation in the survey was that not all people interviewed were of the same generation and therefore even where households had migrated due to partition, often younger generations did not identify themselves as partition refugees. Therefore, it is likely that the actual representation of partition-related migration would be even higher. While partition migration is a form of forced political migration, its relatively large proportion in the sample has warranted a separate category. Partition affected not only Amritsar but also other cities in Punjab, Bengal and other regions where masses of people migrated from their native homes according to the newly demarcated nation-states. Roy (1983) notes that in Calcutta in the Indian state of West Bengal nearly 2.4 million partition refugees settled in the Calcutta Metropolitan District at partition forming a new system of housing access through the legal refugee colonies. A study of Lahore, while not estimating the partition refugee population, attributes partition migration in the cities of Karachi, Lahore, Multan and Peshawar as the main thrust towards high levels of urban growth (Ahsan 1988).

The final category *other* consists of several different types of migration: displacement due to the Indo-Pakistan war, relatives, 1984 anti-Sikh riots, education and natural disaster. Because of the low percentages of each of these responses, they have been statistically combined into the single category *other*. As mentioned in the description of place of origin, migrants from border villages have predominantly settled in Amritsar because of the 1965-1971 wars between India and Pakistan. These people were given compensation by means of housing allotments in public housing tenements and financial assistance. Similarly, the floods between 1965-1966 caused the displacement of people from villages along the border for which compensation was given by the government, though rarely delivered to the appropriate victims of displacement. Waryam Singh Colony, which was purpose-built to house these refugees, shows that very few of the actual allotees ever occupied these flats. The anti-Sikh riots in November 1984 resulted in the influx of many Sikhs into Punjab from other parts of India where their personal security had been threatened. Most of these people who settled in Amritsar did so because of relatives, not because of the safety that Amritsar provided. Amritsar, in fact, became the politically most turbulent area for the next decade as a result of the location of the Golden Temple, the separatist movement and subsequent police repression.

A brief look at the permanence of migrant households shows that fewer migrant households expressed their intention of staying permanently than they did of not staying (Table 6.7).

Table 6.7 Permanence Among Migrant and Non-Migrant Households

	Migrants	Non Migrants
Not Permanent	21	11
Permanent	79	89
Number in Sample	113	162

The reasons given for migration are also closely associated with the degree of permanence of the household. Political refugees, apart from partition refugees, show a high number of people not intending to stay permanently while economic migrants and partition refugees reveal a larger proportion of people wanting to stay permanently in their current homes. Capitalist agricultural development in the north west of India, and more specifically Punjab, has led to the intensification of farming practices through mechanisation and other labour-displacing factors which have severely affected the poor (Byres 1981). The increasingly seasonal nature of agricultural employment for the rural poor has required many to migrate to urban areas on a seasonal basis in search of employment opportunities (Crook 1993). While such processes may lead to the conclusion that the

Figure 6.3 Permanence of Stay among Migrant Groups

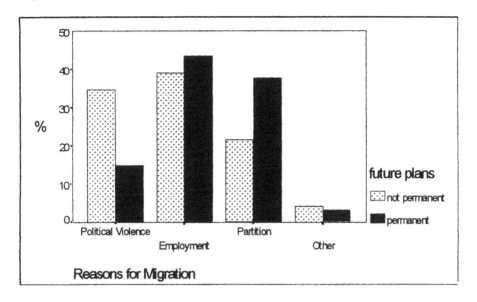

housing settlements of such migrant groups are temporary, when resources and opportunities are available, migrants wish to remain in their new homes on at least a semi-permanent if not permanent basis. For example, Meera Bapat (1981), in her study of hutments in Pune, found that the poor housing quality of hutments was not due to the temporary nature of the migrants' settlement status, but rather because of their inability to afford better housing. Another study by Crook (1993) of India's industrial cities concluded that migrant families preferred to form family units when given the opportunity. A similar finding has been made in the survey in this study where a large proportion of economic migrants have come from eastern regions of India but who also express their intentions to stay in their homes in Amritsar permanently. This point will be revisited in the next chapter when permanence of stay is further explored.

Employment

Sectoral definitions of employment categories are commonly used (such as the terms 'formal' and 'informal,' 'organised' and 'unorganised') in order to describe the security, benefits, wages and further professional prospects of various forms of employment (Desai 1995). Classifications of employment categories are also commonly used in defining employment.

111

Sandhu (1989: 94) in his study of Amritsar's slums uses a grading system by which he divides employment groups among three prestige categories: middle class, lower middle class and lower class.[11] The Census of India data relating to total workers is divided between main workers and marginal workers; main workers being those who have been engaged in some economic activity for six months or more out of the year and marginal workers being those who have worked for some time during the last year but who have worked for less than six months.

Main workers are divided into four general categories: cultivators, agricultural labourers, household industry workers and other workers. Occupations in this study have been classified through generalised categories which act as umbrellas for a range of activities sharing similar conditions. Formal and government sector jobs tend to pay higher wages and offer better long-term security. However, it was noted during the fieldwork that the distinction between formal and informal types of jobs was often unclear and indistinguishable. Though the formal sector may officially be expected to offer a minimum wage and contractual employment, there were many cases in which the same patterns of informal employment were evident in formal employment settings.

The 1991 Census of India altered its conception of 'work' from the 1981 Census by also including unpaid work, mainly the participation in economic activities on farm and family enterprise (Census of India 1991: 16). Inevitably, the inclusion of unpaid work closely related to the economic position of women. However, the data collection in this sample study still faced difficulties in obtaining the correct data on the participation rates of women and children involved in economic activities in the household. This was mainly due to a general lack of recognition that such activities could be considered as forms of employment and in difficulties in recording such activities in the survey.

A number of categories (Table 6.8) have been adopted in describing occupations with examples of common professions found in the survey sample within each.[12] The representation of occupations in the survey are shown in Table 6.9 where the highest number of workers are involved in unskilled/manual activities and the lowest number are involved in craft and professional activities. The category of craft professions includes caste-affiliated artisan professions which have historical attachments with communities and also includes other occupations such as textile printing and musicians which require a certain level of expertise and talent. Skilled occupations include those which in a similar manner to craft, require a certain level of skill but are based upon a degree of training such as the

112

Table 6.8 Employment Categories

1	**craft**	Musician, tailor embroidery, weaver, textile printer, printing press
2	**skilled**	Barber, stenographer, tailor, factory worker, mechanic
3	**government**	Railway, customs, military, municipal corporation sweeper
4	**professional**	Teacher, dentist
5	**trader**	Vegetable/fruit seller, dairy farmer, own shop
6	**unskilled-manual**	Rickshaw, construction, painter, cleaner, domestic servant, shopworker, waiter
7	**marginal**	rag/plastic picking, shoe polisher, ornament/toy making
8	**agricultural labourer**	wage cultivators

Table 6.9 Representation of Occupational Categories

Occupation	Percentage
Unskilled Manual	44.36
Government	18.18
Agricultural Labour	9.45
Skilled Sector	8.36
Marginal	7.27
Trader	7.27
Craft	2.91
Pension	1.45
Professional	0.73
Total	100.00

operative in a factory or in a small-scale private business such as in a tailor shop or a printing press. Some people in this category as self-employed such as tailors and barbers while other such as mechanics, stenographers and factory workers are employed by local businesses. Most skilled workers have acquired their skills as apprentices to established skilled workers or relatives than through any formal training.

Government jobs is a narrow category in the manner that it is used here. Specific attention is paid to low-income groups and therefore automatically excludes medium and high-scale salaried jobs in the government sector. However, even lower-scale salaried jobs show a certain level of diversity of occupations. Railway workers, such as ticket collectors and station attendants, showed to be one of the most secure professions due to long-term security of employment, the potentials for a housing allotment in railway quarters, and the prospects of receiving a regular pension after retirement. Municipal Corporation sweepers are another occupational group whose pay is regular and therefore in demand by poor communities.

As would be anticipated, professionals are the smallest group in the sample with only a marginal number of teachers, dentists and traditional doctors found in the low-income settlements in the survey. The professionals that are represented serve their own local communities and generally do not practice their professions outside of their own vicinity. The level of qualifications that they hold are minimal, often absent, making them rely upon their own personal skills and knowledge rather than any formal training.

Traders are a group which contain a variety of occupational activities. The largest proportion of traders are street vendors who sell vegetables, fruit and cooked food throughout the city. These people roam the streets of the walled city and residential areas in order to sell their goods as well as set up stalls in fruit and vegetable markets. Others in this category have their own shops within their residential area, generally within the home, selling low cost household good to other poor households. Unskilled-manual occupations include such activities as rickshaw pulling to waiting in restaurants. Rickshaw pulling is a low-entry, self-employed profession which requires little capital investment and immediate cash gratification. However, cycle rickshaws can either be rented, leased or bought from dealers which makes the economic position of rickshaw pullers disparate. Some could be categorised as self-employed while others could be more adequately described as commissioned employees of cycle rickshaw businesses. On the other hand, shopworkers and domestic servants are paid by an employer on the basis of their daily labour. While the diversity of occupations within this category are apparent, a generalisation of this category is made on two criteria: the skill required as well as the intensity of manual labour input required.

Marginal workers are those who are employed in the most informal sectors of the economy. Gilbert and Gugler (1981) use the term 'misemployment' when describing the activities in this category. They argue that rural migrants continue to flock to urban areas in order to partake, however little, in whatever demeaning, low-paid tasks available to them. In the sample most marginal workers, as is illustrated in the above table, are *jhuggi* residents and migrant households. This is further illustrated in Figure 6.4, which depicts plastic recycling work in Bangla Basti. This also exemplifies the structurally marginal position that migrants occupy in Amritsar's low-paid occupational structure.

Agricultural labourers are those people who are working for wages

114

Figure 6.4 The Plastics Recycling Industry in Bangla Basti

on another person's land. The private self-help settlements which have come up along the outer edges of the walled city's dumping ground lie in close proximity to agricultural tracts of land. While some people have come from rural backgrounds and have experience of working in the fields, a majority of people have never owned their own land upon which to produce agriculture. Therefore, a majority of agricultural workers in the sample, just as unskilled manual labourers, are selling their labour for cash wages. The opportunities for wage employment as agricultural labourers is seasonal and highly discriminatory in the level of wages paid. The average monthly income for agricultural labourers ranges between Rs. 200 to 900 per month; Rs. 200-300 for women and children and Rs. 600 and higher for male workers. While a majority of agricultural labourers are women, these women responded that farm labour was the most difficult job and that they would prefer to work as seamstresses. Because the competition with professional tailors is great, they are forced to take whatever work they can find. Agricultural work is the most abundant, though not regular, type of wage-earning activity available to them.

Income

The extent to which housing improvement, nutrition, education as well as other factors are influenced by income and financial security has been explored by a number of researchers.[13] Wegelin and Chanond (1983) found that in their study of Bangkok the reasons given by slum dwellers for not improving their houses was overwhelmingly due to lack of finances and less so as a result of fear of eviction. With regard to the effects of income upon health and illness suffered by the poor, another study of Aligarh City in India discovered that a close relationship existed between the types of diseases suffered by family members and the level of income of the household (A.L. Singh et al 1996).

Economic and social indicators of poor households, as exhibited in these studies, are directly affected by the amount of income available for expenditure upon reproductive development of the household. The use of the term 'low-income' as a defining term of the poor by many studies of housing for the poor in Third World cities has been widely accepted, though general in its description. A low income in one urban context may be a high income in another context due to differing costs and markets of housing, land and other subsistence expenses in different settings. Therefore, while 'low-income' may not be a universally applicable definition, a categorisation of income levels is worthy of consideration for a study of access such as this one. Monthly incomes of households in the sample have been grouped into five bands.

Table 6.10 Monthly Income

Income Categories	Number within sample	Percentage of Total
1. Under 1000	18	6.4
2. 1000-1999	125	45.4
3. 2000-2999	60	21.9
4. 3000-3999	38	13.7
5. Over 4000	34	13.7
Total	273**	100.0

*Rs. 55 = 1 pound sterling (1995)
** 2 missing cases

The majority of households in the sample earn between Rs. 1000 to 1999 per month while the lowest portion of households earn under Rs. 1000 with nearly all of these cases being in *jhuggi* settlements. Income levels reveal the abilities of households to invest into further reproductive development. No data or information is available on the distribution of households by income at the city-wide level in India. Therefore, comparison to studies of other cities is useful in developing a means to discuss the economic position of low-income communities in Amritsar.

A research study conducted by the National Institute of Urban Affairs on the urban poor in twenty different urban centres in India found that the incomes of workers varied a great deal depending upon the hours and days worked, occupation and economic sector (NIUA 1993: 40). Over half of the workers in the study earned between Rs. 200 and 600. Another study of slum upgrading in Anna Nagar in Madras found that the average household income was Rs. 638 (de Witt 1992: 41). Datta (1988: 50) found in his study of a medium-sized city, Midnapore in West Bengal, that the median income was Rs. 401 to 600 per month. A more geographically proximate city to Amritsar, Ludhiana, showed an average income of Rs. 401 to 800 in a study of the impacts of the Slum Improvement Programme on the cities low-income settlements (Sandhu 1988: 22). In comparison to these studies of other Indian cities, the mean average income of Amritsar's low-income households is between Rs. 1000-1999, with a median of Rs. 1800 and the mode, or most frequently occurring income level, of Rs. 1200. Both are figures considerably higher than the South Asian cities examined in the other studies.

Different employment activities offer varied degrees of economic security and levels of income. The categories spelt out in the previous section on employment described the types of income generating activities and the income levels and security that they offer. The most prevalent distinction lies between government and marginal sources of employment where government jobs offer security and steady levels of income while marginal work has little security and pay levels fluctuate. Table 6.11 shows the average income levels for each of the employment categories.

Table 6.11 Average Monthly Income

Employment Categories	Under 1000*	1000-1999	2000-2999	3000-3999	Over 4000
Unskilled/manual	9.8	54.9	17.2	10.7	7.4
Trader	5.0	45.0	20.0	10.0	20.0
Skilled	--	56.5	13.0	8.7	21.7
Professional	--	--	50.0	--	50.0
Pension	--	25.0	75.0	--	--
Marginal	10.0	60.0	15.0	15.0	--
Government	--	16.0	30.0	26.0	28.0
Craft	12.5	37.5	25.0	12.5	12.5
Total	6.4	45.4	20.9	13.7	13.7

*Income in Rupees.

There is a high percentage of professionals earning either Rs. 2000-2999 or over Rs. 4000 while the distribution of incomes among government employees is slightly more diverse. Since government employment ranges from sweepers to, in this study, low ranking clerks and officials, the category has been generalised on the basis of job security. As in Chapter Three where the effects of partition and the decline of Amritsar upon labour patterns were presented, the income differentials among skilled, unskilled and marginal workers can be noted in the sample to be relatively comparable. The absence of a large manufacturing sector and an increasingly obsolete crafts and skilled textiles sector has led to the overall decline in wages of such skilled professions. This is significant, as skilled, formal employment has traditionally been viewed as the most secure means of income generation for the urban poor while other professions have less regulated wages and unreliable terms. In contrast to these findings, Desai (1995) found in her study of Bombay's slums that skilled formal sector workers earned on average 79 percent higher incomes than did unskilled informal sector workers showing the positive effects on the poor of organised industrial sectors. Such studies provide means for comparison, though relative in their applications, of low-income groups' positions within other urban contexts.

The types of housing that low-income groups occupy can be related to income levels and the ability to invest into housing costs. Income has commonly been applied by the market approach researchers as a technique for assessing affordability for housing. Singh (1993) comments that the shift from need-based to demand-based analysis in housing studies has emphasised consumptive, investment and speculative motives determined by interest rates, prices and future incomes. Mehta and Mehta (1989) showed in their study of housing affordability in Ahmedabad that contrary to popular opinion, urban poor households spend a higher proportion of their incomes on housing than do other groups with a ratio of housing prices to income estimated at 4 to 3. Struyk (1988) similarly applies income variables as the primary components of influence upon affordability of housing which determine the match between housing needs and solutions. The use of income variables is only applied in a limited capacity so as to assess the economic position of households with regard to the access of housing.

In this chapter a number of social and economic categories have been discussed as indicators of socio-economic status. While it has been noted in this chapter that the heterogeneous nature of poor communities causes difficulties in constructing generalised categories, a number of observations have been made with regard to the economic and occupational profile of the sample households in this study. The significance of religion, caste and migration in the social construction of communities has been addressed in the context of India, Punjab and Amritsar. In Punjab historical cohabitation of different caste and religious communities, the partition as well as capitalist agricultural policies have impacted upon the social make-up of the region. Due to Amritsar's centrality in pre-partition Punjab, the division of Punjab resulted in the erasure of the sizeable Muslim community and in the influx of Hindu and Sikh refugees which is reflected in the high proportion of such refugee communities around the walled city. The more recent phenomenon of economic migrants from poor regions of India is illustrated in the *jhuggi* settlements where recent arrivals find accommodation. Having presented a background to the social and economic indicators used in the study, the next chapter will examine issues around the settlement process and tenure arrangements.

---

[1] Scheduled castes refer to the lowest position in the *varna* or caste system which stratifies and ranks society according to status and profession. The Indian

119

Constitution lists such castes in Article 341. Scheduled castes, also known as untouchables, *harijans* and *dalits*, and have suffered systematic oppression in South Asia for thousands of years. See K.L. Sharma (1994) *Social Stratification and Mobility*, Rawat Publications: New Delhi.

[2] Kshatriyas, Vaishyas and Sudras fall in between the Brahmins and Scheduled Castes. However, for the purposes of this study the caste hierarchy has been simplified and tailored to the predominantly low caste make-up of the survey.

[3] Gill (1991:60) explores the changes caused by urbanisation upon caste and occupational structures in an urbanising villages near Ludhiana, Punjab. Also see Srinivas (1980) who further examines the practice of the caste system by Christians, Muslims and Sikhs.

[4] The Valmiki community follow Rishi Valmiki, a figure from Punjab who is said to have been the author of the traditional Hindu text the *Ramayana* (Saberwal 1990:52).

[5] The Census data on Punjab places Valmikis under the religious category of Hindu.

[6] See Table 3.3 'The Changing Punjab' for a breakdown of the religious composition of pre and post-partition Punjab.

[7] The highest concentration of Muslims in East Punjab are in areas such as Faridkot and Malerkotla.

[8] A study of urbanisation in India notes that one-third of all people living in large Indian cities were born outside of their current city of residence (Nagpaul 1996).

[9] See Figure 3.1 which depicts the political geography of Punjab: Past and Present.

[10] Gugler (1982) argues that at least some of the urban work force could be more efficiently and productively utilised within the rural economy where, though incomes might be lower in rural areas, subsistence and travel costs would also be much lower.

[11] A ranking system which categorises occupations by their prestige or class associations is a highly subjective process, as Sandhu also argues, which has not been attempted in the employment analysis of this study. Therefore, this type of classification, whilst developed specifically for Amritsar, is not appropriate for this study and has not been applied here.

[12] The categories developed here could be seen as a synthesis of the Indian and British Censuses. See Appendix 3.

[13] Sarin's study of Chandigarh's non-plan settlements found that the constraints of economic circumstances upon parents was reflected in the lack of educational and other opportunities available to their children. She notes that the reasons for not sending children for education was not due to the ignorance of uneducated parents, as commonly perceived, but because of economic hardship which required elder children to look after the younger children (1982:136).

7 The Settlement Process

Introduction

The settlement processes by which poor households find shelter show that the path to acquiring housing through either individual illegal occupation and, more so, collective illegal occupation is becoming increasingly narrow. While the scale and severity of the bleak position of the poor in the local housing market in Amritsar does not compare in relative terms to many large cities in either South Asia or in other developing countries, the survey here finds that similar trends to those in other cities are beginning to emerge. Commercialisation pressures upon the area around the walled city are gradually pushing out the most vulnerable groups while also increasing the market values of land and houses in the area. However, the slow rate of legalisation of illegal settlements around the walled city has had the effect of maintaining a high sense of insecurity thus delaying the resale of houses due to the lack of registries. The ineffectiveness of the public sector to promptly legalise illegal settlements has similarly prevented the poor from accessing adequate shelter or improving their existing homes. This chapter will examine the development of low-income settlements in Amritsar and highlight some of the most conspicuous tenure and security issues which confront poor households in shaping their respective communities.

Settlement and Consolidation

Settlement and consolidation of poor communities has been the subject of much discussion within the literature. Experiences of settlement in Latin American cities were among the earliest studies to document the dynamics of the emergence and development of illegal settlements. The detailed analyses of the Latin American context of low-income settlement contrasts with this study of Amritsar since the Latin American experience has a number of features such as the redistribution of agricultural lands and mass organised invasions which have not been prevalent in the South Asian context. However, there is relevance with regard to the concepts of informal and formal, legal and illegal and spontaneous forms of settlement with Latin America acting as an empirical testing ground for the applicability of terminology to the evolving low-income housing markets.[1]

The different actors involved in the low-income settlement processes of illegal colonies in Asian, African and Latin American contexts depict the varying sociological patterns in different Third World cities. The colonial legacy provides the basis for much comparative analysis of settlement processes between African and Asian countries. Colonial legal and administrative frameworks which have, in many countries, been left in tact while the evolution of hybrid systems in other countries which combine the colonial with pre-colonial, traditional systems have further complicated the development of settlement models (King 1990; Lea 1983; Zetter 1984). In Bamako, Mali, for instance, the interaction between African customary land tenure and French colonial tenure resulted in the operation of parallel systems which catered to the European population and at the same time strictly controlled and formalised pre-colonial, collective land rights (van Westen 1990).

A number of studies on South Asian cities have described the settlement process of unregulated low-income settlement (Mitra 1990; Mitra and Nientied 1989). Sarin (1982) traces the relatively recent history of Chandigarh and its non-plannned settlements as being a direct result of the imposition of western planning standards in the city which sought to exclude illegal settlement but which ironically further encouraged the settlement of poor households in non-planned colonies around the periphery of the city. In Karachi the role of middlemen and bureaucrats has been extensively investigated as a main feature of low-income illegal settlement in the city (see Schoorl et al 1983 and Yap 1982) while in Delhi the state's role in monitoring land prices in order to 'socialise' land has had the reverse effect of causing land freezes which has further excluded the poor from accessing legal land (Mitra 1990). Ahsan (1986) describes the *katchi abadis* of Lahore as having been initially formed by partition refugees and that over a period of time migration from rural areas and other small cities began the consolidation process of such settlements.

The parallel history of Lahore and Amritsar in relation to the partition of Punjab shows that Amritsar also has a number of settlements which were initially formed by partition refugees through squatting. In the sample, all of these settlements fall within the category private self-help. *Jhuggi* settlement residents, on the other hand, are more recent arrivals to Amritsar who are predominantly economic migrants from other relatively underdeveloped regions of India and rural areas of Punjab. These settlements, though also having been formed through illegal occupation of land, do not experience the same levels of security due to the increasingly commercialised land use patterns in the city.

There are a number of economic and circumstantial reasons why people choose to live where they do. In the survey respondents were asked the main reason for selecting their present house. The range of elicited responses reflect the multiplicity of factors involved in selection of accommodation (Table 7.1).[2]

Table 7.1 Reasons for Choosing this Site to Live

	Jhuggies	Private Self-Help	State-Assisted	Total Count	Total
Cheap house	--	34	17	68	27
Free land	78	23	--	51	20
Free house	--	18	23	45	18
Near to relatives	22	21	7	44	18
Govt allotment	--	--	30	21	8
Cheap rent	--	2	20	18	7
Near to work	--	2	3	5	2
Total	18	164	70	252	100

Missing Observations: 23
Entries are column percentages and total counts.

Cheap housing was the most common response reflecting the relatively inexpensive costs of purchase of either the house or land and the importance of affordability in the decision-making process of housing selection. However, a distinction can be made between those respondents who chose their house because it was relatively cheap and those who chose their house because the land was cheap. *Cheap house* implies that only the house was purchased with no official purchase of the land, and *cheap land* implies a purchase of the land on which the buyer built their house. *Cheap house and land* denotes those people who purchased the land and ready-built house generally with legal rights to the land. *Jhuggi* respondents gave two responses for settlement: free land and near to relatives. There is no evidence of resale and purchase among jhuggi settlements where squatting on free, vacant land is still the process of settlement. Private self-help settlements, contrastingly, show a higher evidence of resale and purchase. Proximity of relatives was the response of many self-help residents. However, the initial phase of settlement is represented in the squatting on vacant land and the acquisition of vacant houses during the time of partition in 1947.

Free land and *free house* were two separate responses given which indicate either that land was vacant and available for building upon or the house was vacant and was available for squatting. Those respondents who chose to live in their present house due to the proximity of relatives offer an interesting insight into the ways in which housing is accessed among the poor and the role that security plays. State-assisted residents gave more varied responses with 22.9 percent having chosen their home through squatting, showing the inefficiency of government allotment schemes in allocating housing to the targeted populations. This is also due to the fact that a majority of the Municipal Corporation purpose-built flats were not occupied by the original allotees. The attainment of a government allotment only reflects 30 percent of state-assisted respondents' reasons for choosing their current house while 20 percent gave *cheap rent* as the primary reason for choosing this site to live.

The relationship between the location and choice of housing offers another interesting dimension to the analysis of settlement development. There are a number of factors which dictate the household's decision to occupy a house, as illustrated in Table 7.1. The extent to which location determines the selection of housing has an impact upon the spatial and social patterns of settlement. Bradnock (1984) compares Madras' *cheris* to many western city slums in the fact that the importance of being close to places of work is shared by both types of slums. Shakur (1988) in his study of squatters in Dhaka found that city squatters would travel longer distances when the possibilities for higher income existed while camp dwellers had to travel in order to find any kind of work, regardless of income. The city squatters in the Dhaka study can be likened to the private self-help settlements in this Amritsar study while camp dwellers can be compared to the *jhuggi* dwellers. The data extracted from this sample shows only a marginal percentage of respondents showing that they had moved to their current accommodation primarily due to its proximity to their place of work.

Despite the fact that all settlements are found in Zone 2[3] which is close to the economic centre of Amritsar, many respondents expressed their willingness to move further away from the walled city if given the opportunity to acquire better housing. Manjit Singh, a resident of Angarh who works as a tailor in the walled city, explains:

We came here (at partition), not by our own choice, but because we had no other place to go. Now we have been stuck here for nearly fifty years. My children and my children's children have all been born in this house... Of course we would move if someone told us that we could have a house with registry and clean water! But we can't afford to buy a house like that. It would be too expensive.

The location and supply of land for the construction of public housing schemes provides a number of obstacles, one of which being that the largest and cheapest tracts of land exist on the urban periphery and therefore further away from the core of the city (Zetter 1984). Sinha (1991) found in her participant observation study of state-aided housing projects in Lucknow that the location of such projects on the peripheries of the city removed the poor from sources of employment and social ties. Few state-assisted respondents in this study implied that proximity to work was a factor in their housing location selection which illustrates this point further. Usha Kumar, a resident of Ajnala Road comments:

> We moved here to escape from the old city. It is so crowded and dirty there. Some of our family still live in the house there, but we will never go back....Sure, my husband has to travel about three miles everyday to work on his bicycle, but we are happier here. The children have a place to play and we can at least afford to own our own house.

From this it can be inferred that, in general, quality and affordability of housing is a higher priority among the poor households in Amritsar, rather than costs and time involved in travel to and from work. A partial explanation for this can be given in the size and spatial layout of Amritsar whereby most economic activity in the city exists in the highly concentrated area in and around the walled city. Most of the selected sample settlements fall within Zone 2 and therefore can be described as settlements which have emerged in their existing locations due to their relative accessibility to the walled city.

Acquiring a Home

The ways in which households have acquired their present accommodation offers insights into the overall housing pattern. However, illegal settlement patterns show the more discreet processes by which poor households obtain accommodation for themselves thus revealing more longitudinal aspects to housing access. The sample shows a number of different processes of housing acquisition (See Figure 7.1).

Illegal occupation is the single most common form of settlement forming close to half of the total sample across the typology. These residents are settled on land to which they have no legal title or agreement from the owner. The distinction between illegal collective occupation and illegal individual occupation could not be made clear from the sample since many partition refugees settled in their present homes collectively while most non-partition settlers occupied their homes individually through information from friends and family. While illegal settlement has been grouped together in the sample data, there is evidence that collective settlement took place at the time of partition but that individual settlement became predominant as land regulation and zoning laws became more strict in post-independence Amritsar.

Figure 7.1 Means of Acquiring a Home

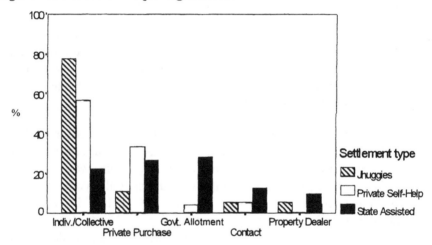

The dominance of Latin America in the urban low-income housing literature has presented much scope for comparing illegal collective occupations with illegal individual occupations in other Third World contexts. Gilbert and Ward (1985: 98) found that in Valencia, Venezuela large-scale invasions had generally been tolerated by official bodies who tended to "take the side of *barrio* inhabitants," though not totally uncontested. In Mexico City they noted that, like other Latin American cities, many squatter settlements formed by rapid invasion which were highly planned and organised with leaders having strategically mapped out the area in advance (Gilbert and Ward 1985). However, they also found that gradual occupation of areas where the land ownership was unclear was also another path to squatter settlement formation, though illegal

126

occupation of private land was rare due to the risks of eviction and the abundance of public land (Gilbert and Ward 1982). Peattie (1982) observed that in Bogota many invasions had been organised by Provivienda, a housing organisation linked to communist activist groups, whose objectives were to provide housing to poor communities through the acquisition of unutilised land. These types of invasions described in Lima by Turner in the 1960s are increasingly becoming rare in Latin America whereas secondary invasions and smaller scale invasions are becoming more prevalent as land controls become more stringent and police enforcement becomes more heavy-handed.

This type of settlement is prevalent in Amritsar as has also been found in other South Asian cities. Individual illegal occupation of land by low-income communities occurs through the gradual settlement of individuals or groups of households. In Dhaka the government's neglect of low-income communities and its patronage of civil servants through subsidised housing schemes has "given rise to a private speculative market where the majority of the population are being either exploited or priced out, and are forced to crowd in slums or squat on vacant public lands" (Shakur 1988: 54). In several Indian cities the enforcement of the Urban Land (Ceiling and Regulation) Act (ULCRA) in bringing urban land on to the market as well as other land policies have had the effect of increasing land prices and thus further excluding the poor (Misra 1990; Raj 1990).[4] In Delhi the process of regularising unauthorised colonies by the Delhi Development Authority (DDA) has led to the further expansion of illegal settlement through subdivisions and an increasingly commercialising land market (Mitra 1990).

The dominant form of informal settlement in Karachi has shifted over the past few decades from mass invasions to illegal subdivisions due to the heightened role of private developers in the public land market (Nientied and van der Linden 1990). In Amritsar a similar shift has taken place from mass invasions which had occurred during and just after partition to individual invasions of public land. Unlike Karachi, there is little evidence of illegal subdivisions by private developers. This is due to two main factors: tighter controls on land, though not as a result of the ULCRA, by both private and public owners than had previously been exerted and the complications of consolidating land owned by a number of public owners. The land around the walled city is largely publicly owned. Over the past fifty years the speculated commercial value of this land has multiplied. Whereas the various public agencies had previously not been concerned about the squatter settlements which had emerged on these areas of land, except for the Waqf Board, they are now becoming more interested in realising their assets through joint ventures with commercial interests. Therefore, the path to acquiring housing through either individual illegal occupation, and more so collective illegal occupation, is becoming increasingly narrow.

Households which purchased their homes from private interests comprise nearly one-third of the sample. This category contains a number of different types of private interests. Private owners who resell their homes is one form of settlement. While there is no significant evidence of resale through property dealers (*dalaal*) in the sample, the rental sector shows a presence of middlemen with only a small fraction of the sample having found their present rented housing through a property dealer to whom is paid at least one month's rent for their services. However, 7 percent of the sample acquired their rented housing through contacts who were familiar to them and to whom a fee was not paid.

Those respondents who purchased their homes from a private interest did not necessarily receive legal title to the land. Often, this only included the housing structure and materials. In such cases, only para-legal rights to the land are included in the purchase. This type of occupation does not exclude illegal settlement; where there is resale houses and legal tenure has not been established. However, it can be inferred from the sample that as a result of low levels of legal ownership, reselling of plots is not a significant segment of the housing market. There is evidence of a direct relationship in other cities between aquisition of registries, increased values and resale,[5] though this does not seem to be the case in Amritsar. However, as more households particularly in self-help settlements are granted registries through legalisation schemes, it seems inevitable that both the rental markets and resale markets will become increasingly important sectors in providing accessible housing to the poor in Amritsar.

Housing and Land Tenure

The relationship between household residents and the house in which they are living refers to housing tenure status. Owners are identified as those people who do not pay rent to any third party. Renters are identified by those people who are paying rent, irregularly or regularly, to a landlord or any other person or organisation. Of the households interviewed, a majority of households owned the accommodation in which they were living (Table 7.2). Further, nearly two-thirds of all renters live in state-assisted accommodation while about one-fourth of all state-assisted housing is rented (Table 7.3).

Table 7.2 Housing Tenure Status

	Jhuggies	Private Self-Help	State-Assisted	Total Count	Total Percent
Own	89	94	76	246	90
Rent	11	6	24	29	10
Total	18	186	71	275	100

Table 7.3 Renters

Jhuggies	Private Self-Help	State-Assisted	Total Count	Percent of Total Sample
11	5	23	29	10

These figures on their own are telling of the commercialisation processes which have crept into the state-assisted schemes. *Jhuggi* settlements, on the other hand, reveal only a fractional representation of renting and show Amritsar to be in stark contrast to other larger cities such as Nairobi[6] (Amis 1988), Karachi[7] (Wahab 1991) and Bombay[8] (Desai and Pillai 1991) where commercialisation processes have infiltrated into extra-legal, traditionally non-commercial housing systems. In Amritsar such settlements still maintain their non-commercial characteristics, here renters comprising 10.5 percent of the sample. The survey findings more resemble Sarin's (1982:160) observations of Chandigarh's non-plan settlements where 13 percent were tenants reflecting a small number of households using their homes as a commodity. In the case of Amritsar's low-income settlements commercialisation processes have been halted by a considerably long period of settlement and consolidation on land owned by public agencies.

Ward (1976) in a study of three settlements in Mexico City identified an increase in owner-occupancy in older settlements which resulted in an increased sense of security. The correlation between the duration of settlements' existence and tenure is not clear in this study as a number of other dependent factors make the correlation slightly less simplistic. For example, the first residents of Bangla Basti and Indira Colony both settled in 1975. Bangla Basti is classified in the typology as a *jhuggi* settlement while Indira Colony is classified as a private self-help settlement. The main distinguishing characteristics of each are that Bangla Basti is mainly comprised of migrant households from eastern parts of India while Indira Colony is primarily made up of local residents. Bangla Basti residents have historically been victims of regular police harassment while Indira Colony has been included in upgrading schemes. Both exist on Municipal Corporation land, though the levels of insecurity and the demographic make-up of residents are contrastingly different. This analysis of renters and owners alone, however, is not sufficient in drawing conclusions about tenure patterns. An examination of land and housing tenure patterns together is a more effective way of assessing tenure patterns within the sample, as will be discussed in the next section.

The rents being paid by renters in the sample can be interpreted through the typology. While the representation of renters in the sample is low, a gradation of rental values can be detected within this group (Table 7.4).

Table 7.4 Monthly Rent

(Rs.per month)*	Jhuggies	Private Self-Help	State-Assisted	Total Count	Total Percent
200-300	100	70	22	11	52
301-400	--	30	22	5	24
401-500	--	--	33	3	14
501 and above	--	--	22	2	10
Total	10	47	43	21	100

Entries are column percentages, Missing Values: 8, *£1= Rs. 55 (1995)

There is a broad range of monthly rent prices in state-assisted housing with rents varying from Rs. 200 per month to Rs. 500 and above per month. The private self-help and state-assisted settlements share similar representations within the sample of renting households. However, given the comparatively low proportion of state-assisted settlements which have been surveyed in the overall study, the relative presence of renting in state-assisted settlements is notably high. The landlords of such rented accommodation are predominantly original allotees who have moved to locations closer to the city where they can use the rental income to subsidise the purchase of other houses. The malfunctioning water and sewerage facilities in Ajnala Road and Gwal Mandi, in particular, have encouraged many allotees to find better housing and facilities elsewhere in private settlements while keeping their allotment as a steady source of supplementary income. Similarly, the Chandigarh case study showed that those people allotted authorised plots, despite regulations against resale or subletting, have been subletting their houses contributing to the steadily increasing incidence of subletting in correlation with the city's commercial housing market (Sarin 1982).

Personal contacts and informal information structures play a significant role in the way that information about housing is accessed. Renting of property between relatives and friends is the most common tenure relationship of those who are renting their accommodation.[9] Though private interests (which include private owners and property dealers) are active in the low-income housing market in Amritsar, their influence in the low-income selling market is minimal and even more so in the low-income rental market. The housing tenure patterns which have emerged from the survey show that the rental sector is small, but gradually growing due to the evolving nature of private self-help settlements and the development of state-assisted sites. Historically in Amritsar, squatting has provided housing to those who could not afford the market prices to purchase land or houses. Due to the increased commercialisation of the area around the walled city, one factor being the Galiara Scheme, land values are becoming more affected by speculation. The impacts of this upon low-income communities living in this area are that squatting is increasingly becoming difficult with heavy-handed police 'protection' of commercial interests. When the various public sector agencies who share the land ownership around the area had generally taken a benevolent approach towards squatters. The land had provided housing to the poor before partition and after partition to the many refugees who had arrived from Pakistan. However, with the gradual demise of squatting as a viable housing option, it is probable that renting will emerge as another path to housing for low-income communities.

Land ownership, escalating market prices and tenure patterns in urban areas illustrate the increasing gap between the masses and the land-owning elites. The structural components of land ownership and distribution such as registration, tax systems and bureaucratic procedures reflect the way in which land is owned and accessed. The common perception by the poor is that private landowners are the main factor in preventing their access to housing (Zetter 1984). However, where policies to curb the concentration of private ownership have been enacted in a number of countries, their success in redirecting land supply to the poor has been negligible. In India the Urban Land Ceiling and Regulation Act (1975) created an official umbrella of 1500 square metres of vacant urban land (Misra 1994:198). However, shadow transfers of holdings and the illegal subdivision of land permitted many to escape the restrictions set by the law (Zetter 1984; Pugh 1990). The resulting fragmentation of holdings made the task of consolidating holdings for public acquisition a difficult task leading some to view the land market as too complex and fundamentally in conflict with wider social welfare goals (Raj 1990).

Empirical evidence shows that land markets in Indian cities are increasingly becoming less favourable to the poor.[10] The significance of land, therefore, is central to tenure as it commands the dynamics of control and access. Land, while influencing the value and affordability of housing, also juxtaposes the problematic nature of private property, which is continually justified within conventional liberal social and economic theory (Pugh 1990). In relation to the access of low-income communities to land for housing, the markets created for land through capitalist development further exclude those who cannot afford market prices from accessing land to house themselves. Through this exclusion, the poor are often left with no other choice than to illegally squat upon public or private land.

Illegality and Legalisation

Legalisation as a policy intended to include rather than exclude the poor living in squatter settlements has also shown to have effects upon mobility. The city of Karachi in Pakistan underwent significant schemes to legalise squatter settlements where it was found that increased market values for the newly legalised areas in the city have seen an insurgence of middle income groups who attempt to enter the settlement by 'pressurising' the inhabitants (Nientied et al 1982). This process operates in a number of ways, but the underlying dynamics are that external forces have differential effects on different social groupings (See Figure 7.2).

Figure 7.2 Reaction to Structural Forces by Social Classes

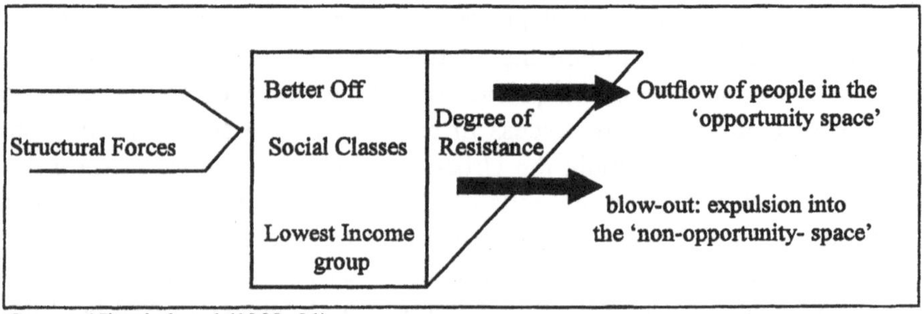

Source: Nientied et al (1982: 21).

In this illustration of the experience of Karachi's *katchi abadis*, the

132

temptation for poor households in need of cash to sell their legalised plots to middle income buyers is shown in the 'blow-out expulsion' whereby the lowest income groups may be forced to settle in a more peripheral part of the city. This model provides a perhaps futuristic view of possible trends in Amritsar's low-income housing system since legalisation has not affected a considerable amount of illegal settlements. While a number of studies have shown a direct link between the legalisation of self-help settlements and their resale and even expansion of the rental market (Nientied et al 1982; van der Linden 1986; Moser 1982) this survey shows that such trends are not yet present in Amritsar. Even where houses have been sold upon receipt of legal tenure, it is low-income groups who have been the purchasers, not middle income groups as the model shown above suggests. The overall small proportion of legalised self-help settlements in Amritsar is reflected in the relatively low percentage of households in the sample that have purchased their houses but who also possess legal rights to the land. Similarly, it can also be argued that the low percentage of resale is at least partially due to the high levels of insecurity due to lack of registries. The slow rate of legalisation by public agencies has postponed such processes of expulsion of low-income groups and the entrance of middle income groups.

The legalisation of illegal settlements by the state and its local authorities has also affected tenure patterns. While self-help, by definition, has sought to emphasise the individual and community in the housing process, the role of the state has also been a central factor. The execution of policies has been managed by local and national government bodies according to the guidelines established by local politics, housing and land markets as well as within the financial and administrative constraints. The resulting impacts of legalisation and upgrading upon the dynamics of social access and distribution have varied. Gilbert and Ward (1985) found that in Mexico City the regularisation of *ejidal*[11] land for squatter settlement improvement generally encouraged further illegal settlement.[12] Rakodi (1995) observed that in Harare land has been acquired by the public sector for low-income housing on an *ad hoc* basis resulting in a slow pace of housing development and installation of services. She argues that the official policy of lessening the gap between low and high density areas through quality construction for low-income settlers has further marginalised the poorest groups through the demand for such housing by middle income groups.

Experiences of regularisation in several South Asian cities also reveal a number of diverse outcomes. The Bustee Improvement Programme (BIP) in Calcutta created a number of complications, one of which was an inflation of rents (Pugh 1990). In Lahore, the Lahore Walled City Upgrading Project[13] beginning in 1982 showed that property values were appreciated and that residents responded positively to the increased property values, although the long-term effects upon socio-economic mobility and access have yet to be assessed (Ahmed 1986). In Delhi the periodic regularisation and upgrading of illegal settlements has resulted in an increased growth of illegal settlements rather than in an improvement of already existing settlements (Mitra 1990). Similar to these experiences, in Amritsar the regularisation of several private self-help settlements, though not in their entirety, has led to the rapid expansion of the size of several settlements. This has often taken place as a result of the residents of legalised land informing their friends and relatives of the increased sense of security endowed to the area resulting in in-migration to these settlements from other low-income settlements. In the cases of Lahori Gate and Angarh, both settlements have been only partially legalised (meaning only a portion of each settlement has been legalised).

Figure 7.3 The Dividing Street Between 'Legal' and 'Illegal' Settlement in Angarh

This partial approach is due to other public and private landowners exerting a vested interest in the land remaining illegal for low-income settlement. As soon as such settlements become legalised, their potential for commercial development overtakes the legal rights of low-income residents to the land. Such vested interests exist despite the fact that many people have been residing in their homes since partition in 1947.

Illegal subdivisions are not a main provider of housing to the poor in Amritsar. However, the gradual sale of public land to commercial developers and the development of illegal housing settlements has resulted in an increase in the privatisation of land. This may in the near future become a heightened trend. Meanwhile, the land tenure patterns in Amritsar show a distinction between house owners and land owners. Because of the concentration of public land ownership around the walled city, the possibilities for most households to gain legal status largely depends upon the granting of legal tenure by the appropriate public agencies such as the Municipal Corporation and the Waqf Board. While many people who identified themselves as owners in the survey were not paying rent to a landlord, the relationship with the accommodation was often ambiguous as a result of land ownership and possession of land title. Renters often were not aware of the tenure status of the land and, as a result, renters have been separated from the analysis of land tenure. Therefore, distinction is made between house ownership and land ownership (Table 7.5).

Table 7.5 Land Tenure

	Jhuggies	Private Self-Help	State-assisted	Total Count	Total Percentage
own house and land	--	39	3	75	27
own house only	89	56	75	173	63
Rent	11	5	22	27	10
Total	18	186	71	275	100

Entries are column percentages and row total counts and percentages.

In the survey, private self-help households have the highest representation of legal status. *Jhuggi* households, due to their highly insecure status, show a high incidence of illegal squatting while 11 percent are tenants. State-assisted households show the highest presence of renters with 22.5 percent identifying themselves are tenants. State-assisted settlements are built upon

135

government land to which individual households do not possess titles, but leases. Therefore, the interpretation of *own house only* in this context implies the possession of a lease granted by the appropriate public authority while *own house and land* indicates those households who hold official documentation to their legal status.

Security of tenure and the granting of legal title are central to many national housing policies for the urban poor. However, misplaced conceptions of security of tenure as a simple concept which merely requires the granting have often led to further complications.[14] In a study of thirteen Bangkok slums Wegelin and Chandond (1983) note that security of tenure as it is perceived by residents is only partially dependent upon their legal tenure status. In this study of Amritsar legal ownership has been examined by using data collected through the survey. Renters have been separated within this analysis of ownership since many renters were not aware of the tenure status of the house in which they were living and could not answer on behalf of the landlord. However, the possession of land titles is one important indicator of the legality of a house and a settlement. In the survey all owners as well as renters were asked as to whether or not there was possession of registry for the land. Within the sample 35 percent actually possessed registries to their homes while 65 percent did not (Table 7.6).

Table 7.6 Possession of Registry

	Jhuggies	Private Self-Help	State-Assisted	Total Count	Total Percent
Without Registry	100	60	68	177	65
With Registry	--	40	32	98	35
Total	18	186	71	275	100

Entries are column percentages total row counts and percentages.

Land tenure is a crucial factor in assessing tenure. In the sample, one-third of respondents (or their landlords) owned the land on which they were living while 65 percent did not own the land. This exemplifies the local government's reluctance in legalising land tenure for self-help settlements and also in its inefficiency in granting leases to public housing allotees. In private self-help settlements 59.7 percent of households do not own the land on which they are living. Similarly, the total sample reveals that a majority of households do not hold legal title to the land.

An interesting, though contrasting, case study of Agra shows how a a private sector real estate development firm acquired and developed land for a low-income housing site resulting in both the granting of land titles to residents and investment returns to the company (Garg 1990). In Bombay Pugh (1990) found that squatters who have settled on private land utilise the services of private developers who arrange for the allocation of plots through established rental agreements and protection payments. In Bangkok private land owners sometimes permit squatters to occupy undeveloped tracts of land on the condition that their tenure status remains temporary (Wegelin and Chandond 1983). The subdivision of plots by landowners and village headsmen in Bangkok has over the years provided a traditional form of low-income access to housing, a system which has recently become threatened by the introduction of strict building and environmental regulations (Angel and Ponchokchai 1990). In the metropolitan area of Karachi eighty percent of the land is owned by either national, provincial or local government, and therefore the granting of legal tenure is largely dependent upon official policy implementation to regularise settlements (Yap 1982).

The spatial dimensions to land ownership patterns in Amritsar are exhibited through the dominance of several public sector agencies which offers grounds for comparison with the Karachi case. The prevalence of the various public agencies as land owners around the city (See Chapter Five, Figure 5.2) has meant that the granting of registries has been dependent upon each agency's policies and interests with regard to the development of its land stock. The fact that most of the low-income households in this sample are living on Municipal Corporation land also signifies that Municipal Corporation land is the most accessible for the poor to build their houses upon while the relatively higher level of security in settlements where the Waqf Board owns the land also reflects the nature of the Waqf Board's activities.

Security from Eviction

One of the central issues discussed at the United Nations Conference on Human Settlements in 1976 in Vancouver as well as at the Habitat II Conference in 1996 in Istanbul was that of the right to housing. The first Habitat conference established the right to human settlement in its charter, though a number of governments, namely the United States, refused to include this in their respective national constitutions. The discussions around the right to housing at the Habitat II conference took the debate a

137

step further to that as a basic human right. Again, twenty years later the motion did not come without opposition and ultimately the decision to establish housing as an official human right became lost in the subtleties of the wording of the resolution and left to the whims of individual governments to execute.

Forced evictions have remained on the agendas of housing rights activists and housing researchers for the past three decades. While the large-scale evictions of the 1960s and 1970s as part of national development programs are generally not prevalent, the pressures of speculation and commercialisation processes on both private and public land have created further complications to the security of low-income communities living on tenuous legal circumstances. The financial gains of developing urban land have encouraged evictions of poor communities simultaneous to the international pressures of human rights and the right to housing. While the right to housing is explicit in international law and increasingly becoming part of the constitutions of national governments, few governments act upon it. Local governments and municipalities are left with few options in implementing the legal rights to housing with pressures from both central governments and private developers (Audefroy 1994).

A study of Bangkok examined the factors affecting the susceptibility to eviction of informal and illegal settlements (Khan 1994). It was found that eviction from public and private land is generally caused by developmental pressures that require land uses which give higher returns on property investment. In Rio de Janeiro the collusion of the local authority with private developers in evicting poor communities has been deemed an "unfortunate but necessary" part of development by both the state and private interests (Kothari 1994). The prevalence of private interests in Third World city land markets is also exhibited in Karachi where, despite efforts by the Karachi Development Corporation (KDC) to find more viable housing solutions for the urban poor through regularisation and upgrading schemes, poor *katchi abadi* residents have continued to be victims of violent eviction and police harassment (Fernandes 1994). Land market values and the aesthetic beauty of affluent areas are threatened by the presence of *katchi abadis*.

In Amritsar the levels of insecurity felt by low-income settlements varies depending upon each settlement's particular circumstances. The question of threat of eviction or demolition cannot be precisely illustrated through quantitative figures alone. Nonetheless, respondents in this survey were asked whether they had been harassed at any time by the police, public officials, bureaucrats, commercial interests or landlords (Table 7.7).

Table 7.7 Threat of Eviction

	Jhuggies	Private Self-Help	State-Assisted	Total Count	Total Percent
Yes	100	39	25	108	39
No	--	61	75	167	61
Total	18	186	71	275	100

Entries are column percentages and total row counts and percentages.

The unanimous response by *jhuggi* residents was that they had been regularly threatened by eviction. Mall Mandi, the least physically developed settlement of the survey, is the most insecure settlement. The land on which the settlement is joined on one side by a private owner with commercial plans for the site and on the other by the Improvement Trust who intends to build flats for middle and higher income groups.

Devdas Ram, a Mall Mandi resident, comments:

> Before coming here, we lived closer to the walled city, near the bus stand. We were evicted at least 5-6 times from other places. Now we are settled here and we don't know how long we will be able to stay here. The owner of the land plans to sell the land or do something with it...he has sent *gundaas* (thugs) a few times to threaten us to get off the land. But where will we go?

Mall Mandi is situated on what had previously been agricultural land but which has increased in value due to its location just off the Grand Trunk road and in the gradual commercial expansion on the eastern side of the city. The threat of demolition is particularly high in *jhuggi* settlements where the occupation of land is blatantly illegal and confrontations with police occurs on a regular basis.

Private self-help settlements also experience insecurity from police harassment. In the survey nearly 40 percent responded that they had been, at some time, threatened with eviction. Many of these households were those which had been threatened during the Emergency in 1975 after which the new government announced that all current squatter settlements on public land should be permitted to stay where they were. This, however, only affected settlements on publicly-owned land whereas private owners in many cases became even less tolerant of squatting.

The sense of insecurity felt by state-assisted housing residents, though proportionately lower than the rest of the typology, is found mainly in the Municipal Corporation purpose-built flats. Here a majority of tenants have squatted in the flats which were meant for a number of different target groups. Legal complaints to the Municipal Corporation by victims of the floods and the Indo-Pakistan war as well as by Municipal Corporation sweepers have led to a number of warnings by Municipal Corporation officials that the current residents immediately leave so that the flats can be appropriately reallocated.

Threats from eviction also influence the household's sense of security. In the household survey respondents were asked whether they intended to stay permanently in their present homes or if they intended to move from their present homes. Such a question is attitudinal in nature and can provide insights into the intentions of households in relation to the physical and tenure conditions in which they are living. The difference of responses among the three types of housing is significant (Figure 7.4).

Figure 7.4 Permanence of Stay

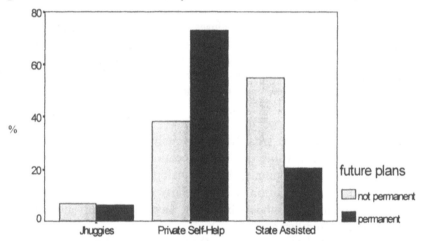

Settlement type

Approximately half of the *jhuggi* residents in the sample said that they were living permanently in their homes while the other half said that they intended to move from their homes either because of threat of eviction or because of their desire to live in better accommodation. As has been presented in this chapter, the infrastructural facilities and location of public housing schemes have been less than adequate in meeting the housing needs of both allotees and non-allotees. The high proportion of non-permanent state-assisted residents are those people who wish to move closer to the city or who feel that they would buy a house in a private residential settlement once they can afford it.

The discrepancy between the responses given by private self-help and state-assisted residents shows that security of tenure is not the single deciding factor (Table 7.8).

Table 7.8 Permanence of Stay According to House and Land Tenure

	Own House and Land	Own House Only	Rent
Not Permanent	5	11	70
Permanent	95	89	30
Total Number in Sample	75	173	27

Entries are column percentages row total counts.

Wegelin and Chanond (1983) found that it was difficult to make conclusions about the difference between the security felt by squatters on publicly-owned land and that felt by squatters on privately-owned land. Similarly, in this study the survey shows that the intentions of residents is formulated on the basis of a number of complex factors which cannot be adequately quantified through data. The intentions of low-income households to stay in their current housing location is dependent on several factors as shown in this survey, one of which is the threat of eviction (Table 7.9).

Table 7.9 Permanence of Stay According to Threat of Eviction

	Threat of Eviction	No Threat of Eviction
Not Permanent	8	19
Permanent	92	80
Number in Sample	108	167

Entries are column percentages.

Despite the absence of threat of eviction, many households (19 percent) responded that they do not intend to stay permanently in their current house while a strikingly high number of people who were under threat of eviction (92 percent) said that they wish to live in their current home permanently. This leads to the conclusion that once security of tenure is obtained, whether official or unofficial, households have a sense of security which allows them to decide their future place of residence. Those households under threat do not have this choice and their future housing is reliant upon gaining a secure status. A more detailed examination of some of these patterns will be further explored in the next and final chapter where the paths to housing available to the poor in Amritsar will be traced and analysed in terms of the opportunities and obstacles that they present.

[1] As Alexander (1988) notes, low-income settlements in Latin American are only informal in the sense that they lie outside of institutional and formal market systems (Alexander 1988).

[2] Respondents were asked to give the single most important motive for choosing to live in their present house. Only one response for each household has been taken into account here.

[3] See Figure 4.1 in Chapter Four.

[4] In Allahabad due to land controls, private owners have sold land to cooperative housing societies who have profited from the buying and selling of land to the highest bidders thus forcing the poor to squat on any available land (Misra 1990:198).

[5] Nientied et al (1982) note the forces, one of which is the increased market values, that push lower income groups out of regularised colonies. In the Dandora Project in Nairobi identity cards were issued to deter the reallocation of titles (Chana 1984).

[6] Amis (1982) argues that the development of a large rental market in squatter settlements is an expected result of wider commercialisation processes within the economy. Thus far, Amis's study shows Nairobi to have experienced the largest transformation of squatter settlements into rental sectors than any other Third World city.

[7] In Karachi about 80 percent of land in the Karachi Metropolitan Area is owned by local, provincial and national government (Yap 1982:29). The legalisation of many squatter areas in the city has had the effect of heightening commercialisation processes which have in turn made renting a more affordable option for many poor households (Wahab 1991:299).

[8] Desai and Pillai's study (1991:83) of Bombay's hutment settlements noted a 70% representation of tenants.

[9] Mathey (1990) revealed in his study of Nicaragua that there were no paramount indications of construction for sale or commercial lease in the private housing sector. However, the sharing of single houses by more than one family and subletting to friends and distant relatives was shown to be on the increase in Nicaragua. Friends and distant relatives in many Latin American countries are referred to, here by Mathey (1990:88) as *inquilenatos*.

[10] Raj (1990:261) notes that the total squatter population in India's urban areas is estimated to be around 30 million and is expected to increase to nearly 80 million in the next fifteen years. Gupta (1985:120) displays evidence that there are approximately 59 percent of households in India who do not own any land at all. This is further exacerbated by 3 to 4 percent of affluent households who own approximately 18 percent of the total land in India.

[11] In Mexico during the nineteenth century most communal land was sold to private owners and then reconstituted as ejidos. Ejidal land is owned by the community but worked on individually by denoted peasants. For a more detailed discussion see Gilbert and Ward (1985) chapter 3.

[12] The dynamics between individualised systems and pre-colonial customary tenure in Papua New Guinea, particularly due to the settlement of migrants, have also shown to have created complications in the public acquisition of land for housing the poor (Lea 1983).

[13] In Lahore residents of illegal *katchi abadis* have lobbied for the regularisation of their settlements through the forming of community organisations such as the People's Planning Project and the Awami Rehaishi Tanzeem (People's Housing Organisation) (Ahsan 1986:82).

[14] See Zetter (1984) for a more detailed discussion of title and tenure. He refers to oversimplification of security of tenure through the granting of title as a result of a 'confusing mixture of traditional and European hybrids.'

8   Paths to Housing Access

Introduction

Social and economic structures such as migration, occupation, class and
caste, reveal the social dimensions to how access becomes differentiated.
Meanwhile, the means of access to housing rely upon the households'
ability to afford the costs of land and built-structures, to access information
regarding public housing scheme allotments and to respond to the
dynamics of the housing market as dictated by both formal and informal
processes.

The theoretical positions on access examined in Chapter Two
outlined several of the main issues involved in the access to housing among
the poor in Third World cities. The departure between the self-help and
neo-marxist positions illustrated the opposing views on the effects of state-
initiated and state-supported actions. The self-help propagators highlight
the merits over conventional housing of self-help sponsored projects as
more efficient uses of limited resources (Turner 1976; Rodell and Skinner
1983; Payne 1984) while critics of self-help sponsored projects deem them
as further marginalising of the poor by exploiting their labour and their
structural position within the capitalist housing system (Harms 1992;
Marcuse 1992; Burgess 1982). The latter forms the premise of this study
which views the development of patterns of access in relation to the
exchange process. Settlement and exchange processes dictate the way in
which households gain access to housing, and this final chapter aims to
illustrate how self-help contributes to the social differentiation of access
through the typology.

The typology developed in Chapter Four will be applied here to
assess patterns of access as the framework of analysis of Amritsar's low-
income housing. The paths to housing for the poor in Amritsar will be
traced in the form of a diagram used to illustrate the prevalent routes by
which different socio-economic groups are acquiring shelter. Next, an
examination of the relationship between social status and access will be
made within the context of Turner's model of mobility. The social
dimensions of the differentiation of housing access in the survey of
Amritsar will then be comparatively analysed to this model.

Paths to Housing Access

Each type of housing occupies a specific place within the low-income housing system in Amritsar. For instance, the predominance of low level employment options, lower levels of income and the comparatively high insecurity of tenure experienced in *jhuggi* settlements reflects their marginal position within the city's social, economic and housing structures. Meanwhile, the range of employment, tenure and income levels in private self-help and state-assisted settlements shows the diversity of social groups. The relationship between social patterns and the types of housing in Amritsar are illustrated in the paths to housing access among the low-income communities. The entry of three groups (new migrants, local urban poor and old migrants) into the low-income housing system of Amritsar is depicted according to the flows of access which surfaced from the survey.

Figure 8.1 illustrates the paths to housing access in Amritsar and presents the flows of housing access according to the socially defined directions of movement in the city's housing system. As can be seen in the diagram, the typology of housing is represented in the solid circles which show the dominant sectors of the low-income housing market. Rental housing, a relatively small and recent addition to the low-income housing market, is not a dominant trend, though the survey in this study found it to be on the increase due to the exclusion of the poor through rising commercial values of housing and land. The solid lines in the diagram show the clearly defined paths by which the poor are accessing housing while the dotted lines in the diagram indicate the less dominant paths. The dotted lines represent residual trends in the case of rental housing state-assisted housing and heightening trends in the case of increases in renting in the commercialising private self-help settlements, a distinction which will be further discussed in this chapter.

Jhuggi settlements have historically provided newcomers to the city with immediate accommodation. Their comparatively low quality of housing and service standards and illegal land tenure make them the least desirable, though the most readily accessible form of housing. Private self-help settlements have emerged from a number of different periods. First, there are those settlements occupied by the local urban poor which existed prior to partition and have become consolidated. Second, partition settlements, in this section referred to as 'old migrants,' came up between 1947 and 1948 due to the influx of refugees coming from newly forming Pakistan. The third period of private self-help settlement development arose out of more recent settlement in the 1970s, particularly during the

145

Emergency between 1975-1977, upon environmentally degenerate, vacant tracts of land, such as dumping grounds.

State-assisted housing in Amritsar has a number of historical stages of development. The first was during the mid-1960s and early 1970s when unserviced Municipal Corporation flats were sporadically constructed for target groups such as Municipal Corporation sweepers, victims of the Indo-Pakistan war and displaced families from floods. These flats lack all basic facilities and are the lowest standard of public housing in Amritsar. The second stage of state-assisted housing development occurred during the early 1980s when the Punjab Housing Board constructed a number of serviced one-bedroom flat tenements for allotment to low-income groups. These flats do not have fully functioning services while their distance from economic activities in the city have disappointed many of the allotees. Due to the increasing pressure upon land immediately around the walled city, the security attached to allotments in public schemes still remains as the highest priority among low-income

Figure 8.1 Paths to Housing Access in Amritsar

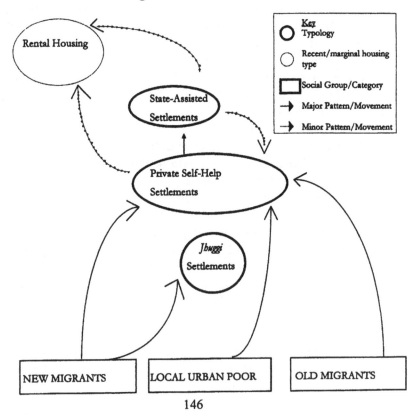

households' preferences. However, the ability of households to access state-assisted allotments is limited to private self-help households who have access to information, who can afford the down payment and who are willing to live at a distance from the walled city.

'New migrants' enter the housing market according to their economic and social status. The poorest migrants, which form the bulk of all new migrants, are those coming from Bihar and Uttar Pradesh. These migrants are drawn toward the walled city where chances for employment are highest. *Jhuggi* settlements around the walled city serve the immediate housing needs of new migrants.[1] There is little evidence of the movement of new migrant households into the walled city as until recently the option to squat or purchase housing has existed in the private self-help settlements around the walled city. However, the increasingly commercialised nature of the activities in the walled city has already seen the elites moving out of the walled city and the poorest groups towards the walled city.[2] A perhaps more significant factor is the comparatively low social status of economic migrants from Bihar, West Bengal and Uttar Pradesh. The marginal position of these groups in Amritsar's social structure is reflected in the spatial segregation of *jhuggi* settlements from even other low-income settlements.

Other 'new migrants' are those people who do not have enough savings to live in better accommodation, generally from other parts of Punjab, but who are able to afford the market prices in private self-help settlements. The 'local urban poor,' who primarily originate from within the walled city, occupied private self-help settlements during the time of partition when Muslim homes located on Waqf Board land were vacated. Partition migrants form the 'old migrants' who came to Amritsar during and after partition and also occupied private self-help settlements. Old migrants who could not find available housing and land at partition began to squat on public land which also subsequently became part of the newly forming private self-help settlements. The movement of households to state-assisted schemes has occurred from private self-help settlements by local urban poor and old migrants. The allotment of flats has by-passed *jhuggi* settlements, recruiting mainly from private self-help settlements where more people are able to access information and afford the initial payment. The reverse movement of households from state-assisted settlements to private self-help settlements is reflected in the growing rental sector where original allotees have chosen to rent their allotments in order to gain a supplementary monthly income as they either move back to the private self-help sector or obtain an allotment in another scheme.

147

The paths to housing access in Amritsar shown in Figure 8.1 trace the entry points and routes by which different social groups position themselves within the low-income housing system. In the next section, a closer examination of the relationship between housing access and mobility will be made in order to draw some conclusions about the patterns of social access in Amritsar's low-income settlements.

Migration, Mobility and Settlement

The relationship between the social status of households and the housing which they occupy has been the subject of much empirical as well as theoretical work. Although the central premise of this thesis is to challenge the self-help school's contributions to housing access, the model presented by Turner (1968 and 1969), the biggest advocate of self-help, of socio-economic position and housing conditions provides a useful comparative tool for analysing access. According to this model the socio-economic status of a household strongly influences the housing conditions and housing paths that are available to them. Van Lindert (1992) offers an interpretation of Turner's model which is used here in relation to the socio-economic variables of access in this study (Table 8.1). This interpretation of Turner's model of settlement gives several insights for the survey analysis in this study as it largely assumes that the socio-economic status of households is the main determinant of their housing access. Turner's model is a highly generalised one which needs to be applied with caution to any specific case study of low-income housing access (Van Lindert 1992). The limitations of such a model are emphasised in the application of categories to specific social groups within other urban contexts. The relevance of the model to this study of Amritsar will be examined in this section.

The socio-economic profile of Amritsar's low-income population in Chapter Six showed that there is a sizeable migrant population (41 percent). All *jhuggi* residents have migrated from outside of Amritsar, one-third of all private self-help are migrants while close to half of state-assisted identified themselves as migrants. While this would show a disparity in the typology as to where migrant households have become absorbed within the housing market, the reasons for migration show that the vast majority of economic migrants from regions in eastern India are living in *jhuggi* settlements while a high number of partition migrants are living in private self-help settlements. State-assisted residents come from a variety of migration histories: partition, economic and political. However, not all households in Amritsar's low-income housing settlements are migrants, with a large proportion being local residents of Amritsar.

Table 8.1 An Interpretation of Turner's Characteristics of Migrant Categories

	Bridgeheaders	Consolidators	Status Seekers
Length of stay in city	less than 5 years	5-10 years	more than 10 years
Phase in family cycle	unmarried	relatively recently wed; child-bearing age	older marriages child-rearing or child-launching stage
Type of employment/ employment status	unskilled casual employment/ small-scale sector	unskilled fixed employment/corporate sector	skilled fixed employment/corporate sector
Type of income	very low, unstable, insecure	low, stable, secure	middle-income, stable secure
Housing priorities	1. location 2. tenure 3. amenity	1. tenure 2. location 3. amenity	1. amenity 2. tenure 3. location
Shelter/habitat	(sub-)tenants in inner-city slum areas	owner-occupants through self-help housing on the present periphery	owner-occupant or tenant, either via complete consolidation of (self-help) housing on the former periphery or in (government-aided) housing schemes

Source: Van Lindert 1992: 159.

The bridgeheaders of Turner's model in this study would, by definition of length of stay, be those newly arrived migrants living in *jhuggi* settlements. However, Turner holds that bridgeheaders prioritise proximity to workplace for settlement selection and therefore prefer inner-city slum rental housing which is cheap and of generally worse condition than housing elsewhere in the city. With regard to the position of bridgeheaders, Turner's model does not hold true in Amritsar. The bridgeheader communities in Amritsar are the newest arrivals to the city and have come from poor, rural backgrounds predominantly from eastern India but also from rural areas in Punjab. Such communities have migrated from between 2 and 20 years. The walled city is only recently becoming a place where the poorest newcomers to the city go to find housing. Up until now, the opportunity to squat on vacant land has existed around the walled city and therefore free housing, for which the possibility of eventual ownership might eventually exist, has served their housing need. The expansion of middle and upper income residential areas in other parts of the city has resulted in the gradual outward movement of upper and middle income households and in the inward movement of lower income households. However, the lack of space and facilities found in much of the rental housing available to low-income tenants inside the walled city does not make it a more favourable or affordable market to the settlements around the walled city.

The length of stay of bridgeheader communities shows only to have impacted on the physical development of settlements where the owner of the land did not pose a threat. Overall, however, bridgeheader communities in Amritsar are stagnant, or immobile. Mall Mandi residents, nomadic squatters for the past 30 years in Amritsar, are continually experiencing threats of eviction and violence. Choice of site for settlement for Mall Mandi residents has been a continual gamble as the unfortunate selection of a plot of land with a potential high market value inevitably results in forced eviction. As squatting has become more difficult, those who have come most recently have become absorbed within *jhuggi* and private self-help settlements by living with relatives and renting. The rental sector in Amritsar, while only a marginal proportion of the low-income housing sample, shows to be servicing newly arrived households.

Turner's consolidators, more than any other group, show similarity with the Amritsar case study. Turner's consolidators are owner-occupier households who are eager to gain legal tenure for themselves. The consolidators in this study would best be described as communities living in more secure *jhuggi* settlements and quasi-legal private self-help settlements. Bangla Basti, whose first settlers arrived in 1975, has a comparatively high level of security to other *jhuggi* settlements as a result of the recent influences of local politicians and in the lobbying of community leaders for services. Indira Colony accommodates families who had previously lived in *jhuggi* settlements of insecure tenure and who have been given unofficial promises by politicians that the settlement will gain legal tenure. Indira Colony is predominantly comprised of young residents who seek to secure permanent homes for themselves. It is these types of settlements which are most noticeably impacting upon the nature of squatting around the walled city where residents are becoming increasingly

150

aware of how to exert their influence through electoral votes and in community organisation (Figure 8.2).

Figure 8.2 The Residents' Association of Indira Colony

As these settlements are granted tenure and serviced through upgrading programmes, the impetus for other households to stake their claims on available land has increased the physical pressure upon the area as a whole.

The status seekers in this study show to be a less homogenous group than as described in Turner's category. This group is represented in both secure private self-help and state-assisted housing settlements where security of tenure or at least security from eviction are positive. The diversity of this category can be most noted as pre-partition residents, partition migrants, government allotees, and Municipal Corporation tenement residents all fall within this category. The migration and settlement histories of each of these groups offers a variety of explanations for their current housing status.

Pre-partition residents such as those living in Lahori Gate have been living in Amritsar since before 1947. Lahori gate was occupied at partition by local walled city residents who saw the opportunity to occupy vacant Muslim homes. In this sense, they are 'status seekers' as they moved out of the walled city to improve their housing conditions away from the congested and overcrowded walled city. Partition migrant communities as those living in Angarh and Gujjarpura settled either in abandoned homes of Muslims or through squatting on public land. These communities have unofficial legal tenure and official legal tenure through the granting of registries and rental agreements from the Waqf Board and the Municipal Corporation. However, facilities are still inadequate as these communities have been unsuccessful in lobbying for attention from political interests. In this respect, consolidators have been more successful in attaining facilities through political patronage than have status seekers. The volatile tenure status of consolidator communities is partial cause of this. Municipal Corporation tenement residents, while not having legal status, are relatively secure in their occupation of the flats. However, the lack of facilities and the quasi-legal nature of their status has made their case for attaining facilities weak. Meanwhile, households in government allotments have full legal tenure. Facilities are not adequately functioning in most schemes, and residents are eager to have improvements made by the appropriate authorities. However, the facilities are not regularly serviced and maintained, and therefore the residents are left to either invest their own money into improvements or wait until the authorities act.

One observation in this categorisation of access is that the legal status of settlements has an impact upon the amount of pressure they are able to exert upon official and unofficial means of housing improvement. Legal settlements have shown to have less ability to mobilise action while outright illegal settlements have been able to exert more pressure. The worst off are those with quasi-legal status whose situation cannot be capitalised upon by political interests and who are most subject to the inertia of public agencies.

In this section, the model presented by Turner of migrant housing mobility has been analysed. Its applicability as a universal template of the dynamics of housing mobility is limited due to the generalised characteristics that it presents. This is due to the ahistorical representation of stages in the households' development which assumes a homogenous progression of bridgeheaders, consolidators and status seekers. The basic concerns of the Turner model are that the individual household's priorities change along with family cycle and length of stay in the city. The model adopted by the wider self-help school fails to acknowledge the structural obstacles that prevent the poor from obtaining secure, conveniently located and adequate housing. Rather, the self-help theorists and neo-classical economic theory in general attempt to show how social status in housing is differentiated through utility maximisation and consumer choice (Harvey 1973: 109). The assessment of housing in this context is an analysis between supply and demand as a market relation rather than as a structural determination. In this study the market imbalance between supply and demand is not upheld as the primary cause for the housing crisis. As will be argued in the following sections on the survey of Amritsar, the dynamics of social access to housing are slightly less simplistic. The movement of different social groups within a particular city's housing system are structurally determined. Disparities in housing access are determined, as Castells (1977: 146) points out, by "needs, socially defined, of the habitat and the production of housing and residential amenities." Though the basic premise in this study, in contrast to Turner's analysis, is that social status is not individually determined, social and economic indicators nonetheless provide a useful method by which to assess the patterns of exclusion and social differentiation of housing access.

Social Differentiation in the Low-income Housing Market

While Turner's model of mobility considers all migrants within its framework of household stages, the experiences of different migrant communities are not singular and are subject to a number of factors which result in unequal relations in the housing system. The low status of *jhuggi* settlements is reflected in their highly insecure and illegal tenure conditions. The large presence of migrants from eastern regions of India in *jhuggi settlements* denotes the marginal position of these communities who are living in the city's lowest standards of housing. The entry point for economic migrants from eastern regions of India is explicitly the *jhuggi* settlements where the least amount of capital investment is required and the security of living among one's own community offers a sense of belonging and solidarity. Segregation of these communities and the discrimination they experience from other communities is also a factor determining the marginalisation of eastern India migrants to *jhuggi* settlements.

The effects of partition upon Amritsar's social patterns of housing are most evident in the high percentage of partition refugee households residing in private self-help settlements. However, due to Amritsar's secondary economic role in Punjab to Ludhiana, large-scale migration is not a feature of Amritsar's demography and the local poor still form the largest proportion of low-income settlements.[3] These households, referred to as 'local urban poor' in Figure 8.1, are concentrated in both private self-help and state-assisted settlements. The local urban poor have shown the ability to take advantage of information which is exhibited in the occupation of abandoned Muslim homes at partition and more recently the accession to state-assisted housing developments.

The privileged status of state-assisted housing settlements in Amritsar's low-income settlements can be illustrated through the social make-up of residents in such schemes. Marxist analysis of social differentiation to housing most commonly uses class to measure differentiation within the social structure (Castells 1977; Harvey 1973). However, here it is argued that pure class analysis of housing is not adequate in the context of this study. The intersection between caste and class is slightly more complex in South Asian urban contexts and requires a cross-examination of class, caste, income and occupation which collectively form the basis of the social structure.[4] Although industrialisation and urbanisation have had effects upon the validity of caste as the primary basis of social structure, as explained in Chapter Six, caste still remains a factor in determining social and economic organisation. It has most significantly continued to mark the living conditions of the poor, bound by occupational and spatial restrictions. In this study, this is most notably evident in the caste and occupational make-up of state-assisted settlements which, in addition to income and employment categories, reveal an overall higher position in comparison to the other types of housing (Figures 8.3 and 8.4). While the aims of state housing initiatives have shown to be generally target-driven in order to service the needs of particular groups, the performance of state-assisted housing schemes in Amritsar reveals how the state's activities have in practice caused further social differentiation. Where housing has been developed for government-identified Low-Income Groups (LIG's) and Economically Weaker Sections (EWS), the survey shows that the poorest and lowest caste groups are proportionately not benefiting from such schemes. Of the high caste respondents in the study 77 percent were living in state-assisted housing while 60 percent of the category 'other,' comprising Sikh middle caste groups were living in state-assisted housing.

This illustrates the caste dimensions of access showing that higher castes are occupying the better housing in state-assisted settlements.

Figure 8.3 Distribution of Caste groups within the Housing Types

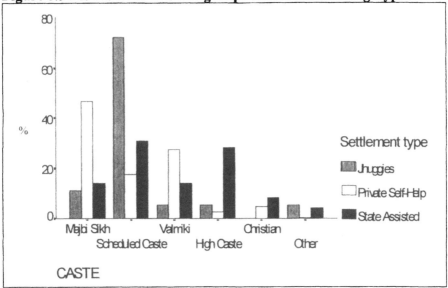

Figure 8.4 Distribution of Employment Categories by Housing Type

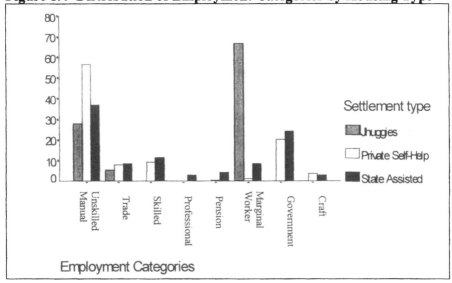

Figure 8.4 illustrates the employment categories which were described in Chapter Six. Marginal workers have the lowest and most irregular pay conditions and are concentrated in *jhuggi* settlements where households cannot afford market prices. Unskilled/manual jobs are distributed throughout the typology with a high representation in private self-help settlements. The higher paid professions of skilled workers, professionals and government workers are concentrated in state-assisted, though also in private self-help settlements. This exemplifies the diversity of the occupational profile of low-income settlements but also shows the correlation between higher status jobs and the type of accommodation that households occupy. State-assisted housing remains at the top of this hierarchy of social status. Allotments in state-assisted schemes are viewed as a priority in the survey by a majority of households, and the malfunctioning services are considered to be more favourable to the absence of services in *jhuggi* and private self-help settlements.

Income, Affordability and Differential Access

Affordability has been used by a number of studies as a main indicator for market analysis of housing access. Mehta and Mehta (1994) examine the question of housing and affordability in the context of India, asserting the need for a market-based approach. The goal of providing comprehensive state housing, they argue, is unaffordable for the public sector and therefore the case for a welfare state is limited. They note that while living conditions in urban areas have deteriorated over the past decade, the access to basic services and housing has also deteriorated with nearly twenty-five percent of the urban population in 1981 without adequate shelter. Their market-based perspective views affordability in relation to demand-based household characteristics: income, social and ethnic characteristics, stage in the life cycle: "...the issue of housing affordability relates to the ability and willingness of the household to pay for housing and the prevailing housing prices or costs" (Mehta and Mehta 1994: 19-20).

Mathey (1992), on the contrary, challenges the reason why people build their own houses in capitalist countries as a cost-saving incentive by not hiring wage labour. He illustrates how in Cuba's microbrigades the price of the house is discounted through the exchange for the self-help labour input. However, a comparison of self-help and state-built/microbrigade flats shows that self-help building does not significantly reduce the costs of housing for the household, and in many cases, may actually increase the cost. Therefore, the 'logic' of self-help building in order to save money should not be an incentive for households to participate in self-help housing.

The amount that households can afford to pay for housing determines, to a large degree, the paths available to them for movement within the housing system. The most common reason given by respondents in this study for their housing situation was the lack of economic resources. Chedi Lal, a resident of Bangla Basti comments:

> We came from Bengal to find work. Here we don't even earn enough to send money back home or to save.... all of our income goes into running the house...there is nothing left at the end of the month for us to save...

Misra and Gupta (1981) state that the cost of 'acceptable' housing in big cities simply exceeds the income power of a large proportion of the population. The costs of making physical improvements or of purchasing another house or plot in a legal, fully serviced area is generally out of reach of urban families who do not have substantial savings. The amount that households can afford to pay for housing is intrinsically linked to the types of employment that are available to them. Access to reliable forms of employment is structurally defined by the labour market and often, in the case of Amritsar, the historical participation of communities in specific activities. This has led to the exclusion of the poorest from the best jobs and in their higher rate of participation in lower status jobs.

The concentration of *jhuggi* households in unskilled/manual and marginal employment activities is significant in the sample. These two categories are the least secure and, as will be discussed with regards to income levels, are the lowest paid occupations in the city. Private self-help and state assisted households have more diverse occupational profiles. However, private self-help households show a higher representation of unskilled/manual workers while the state assisted households have a higher percentage of government workers.[5] There are a variety of employment activities within all three types of housing. The lowest paying and most irregular jobs are overwhelmingly done by people living in the most insecure, poorly constructed and serviced housing. The ability of *jhuggi* households to afford better housing, therefore, is marked by their occupational profile.

Though the levels of income for each category are important for understanding income distribution, the security which each employment category carries cannot be adequately assessed through figures. This can be most distinctly noted with government workers and pensioners who may not necessarily earn higher wages than craftsmen and professionals. However, their long-term earnings and capacity for savings put them at an advantage to other occupations. Income levels, nonetheless, reveal the amount of money that is available for household expenditure. The distribution of income levels within the typology reveals that a hierarchy of average incomes exists between *jhuggies*, private self-help and state-assisted settlements, in ascending order (See Table 8.2).

Table 8.2 Levels of Income

Income Level	*Jhuggi*	Private Self-Help	State-Assisted
Under 1000	28	6	3
1000 to 1999	44*	46*	45
2000 to 2999	17	23	18
3000 to 3999	11	14	14*
Over 4000	--	11	20
Total	18	186	71

*Indicates range of median income.

The median range of income for *jhuggi* and private self-help settlements is between Rs. 1000 and 1999. However, the average household size is higher among *jhuggi* households than in private self-help households, 7 and 5 respectively. The median income for state-assisted households is considerably higher being between Rs. 3000-3999 (Table 8.3).

Table 8.3 Average Income

	Jhuggies	Private Self-Help	State-Assisted	Total
Standard Deviation	733	1300	1838	1452
Mean Income	1452	2137	2545	2197
Total in sample	18	186	71	275

Entries are column calculations.

Both mean and median have been used in analysing income levels among the three types of housing in order to provide an overall picture of the income patterns that are present within each type. The mean income for *jhuggi* settlements is Rs. 1452 while private self-help and state-assisted are Rs. 2137 and 2545, respectively. Within Amritsar's low-income settlements those with higher income earnings are disproportionately represented in state-assisted housing for which only a small sum of cash payment is required as a deposit and the monthly instalments are comparatively low. As has already been noted in Chapter Seven that commercialisation of housing processes are yet at their infant stages in Amritsar, the housing prices in private self-help range from Rs.10,000 for a 30 square yard plot to Rs.100,000 for a built-house on a 50 square yard plot. The price of a house depends upon the legal tenure of the land most specifically, but also upon informal tenure arrangements, services and location. The poorest are left to reside in illegal *jhuggi* settlements while those with access to savings or capital are able to enter the housing market. However, the connection between payment for housing and the improved quality of housing is not a direct one. The survey reveals that while those who are purchasing houses are doing so with savings, this is due to the increasingly competitive and commodified housing market rather than in an improved, and therefore for more expensive, housing stock.

The recent activities of public sector agencies in the legalisation of private self-help settlements as well as in the construction housing schemes have begun to alter the nature of the low-income housing system in Amritsar. Legalisation of private self-help settlements, though not in their entirety and most particularly of partition settlements, has resulted in the introduction of commodification processes in such settlements. Where previously market prices for land and houses in illegal self-help settlements could not be applied, there is evidence that the recent granting of titles to a number of households in such settlements is resulting in the integration of low-income settlements into the land and housing markets. Where land and housing in *jhuggi* and private self-help settlements had once been provided to low-income households through non-commercial means such as squatting, the ability to find free housing in Amritsar is becoming increasingly difficult. The representation of households who have purchased their homes from private interests in state-assisted and private self-help settlements is representative of such commodification processes. A majority of reselling is a result of legal tenure, though there is evidence that even those houses without registries have also been bought and sold. This is most evident in the small proportion of *jhuggi* households who have purchased their homes without legal status.

Bhajan Das, a resident of Gujjarpura comments:

...it is very difficult to get an allotment in a public scheme. All that my family wants is to know that we can live in our house safely without being bothered by anyone. Even though we are still waiting to get a registry for our house, we know that we will eventually get it. We have begun to improve our house with a hand pump and a toilet..... but if one of my sons can get an allotment in a public scheme, I would definitely want him to move. We should have some security....

Another aspect of affordability is the ability to pay for services. Respondents were asked if they would pay for services if they were asked to do so. The overwhelming response was positive. As one person, a resident of India Colony comments: "Yes, we would pay for services, that is, if they were ever delivered. The problem is not in paying. The government has to decide that it wants to do something..." Van der Linden (1983: 47) notes in Karachi that residents are even eager to pay for facilities as they, "besides serving their intended function, enhance the security level of the *basti*." The correlation between the demand for services and an increased sense of security from service provision is also explicit in this survey.

Renting is one aspect of the housing market which will be further discussed in the next section of this chapter. Owning has been a cheaper form of accommodation than renting in Amritsar since there has historically been free land available for the poor as a result of the active involvement of politicians in illegal occupations and the lack of strict controls on land use. However, those households who occupy insecure land lacking services tend to be the poorest in the city. These households are *jhuggi* and poorer private self-help residents who lack the economic resources to invest into better housing development. In a similar case, Sinha (1991: 20) notes on the exclusion of the poorest groups from state-supported housing in Lucknow, which also holds true in the case of this survey in Amritsar, asserting that the "institutional structure of initial down payment, regular monthly payments and the penalties attached to default, does not fit in with the irregular income of construction labourers, street vendors and others in typical marginal occupations."

Those households who have enough earning members or who have enough savings to buy land or houses in secure and serviced areas have more options to improve their housing. It is these households who have specifically benefited from state-assisted housing schemes and upgrading schemes. Despite the fact that allotments for low-income groups in state-assisted schemes require small amounts for deposit (on an average Rs.10,000), the ability to pay the money up front and, in the case of legalisation and upgrading, to pay government-associated officials or other less official interests involved in such procedures heightens the chances for housing improvement. Therefore, affordability not only relates to the monetary power that a household can exert in the housing and land markets, but also in the level of confidence with which a household can go

160

about gaining information and making contacts.

Legality, Tenure and Access

A commonly noted point within the low-income housing literature is that of the relationship between land tenure and housing improvement. The rationale for the granting of land tenure as most popularly presented by Turner (1976) is partially based upon the logic that people, if given a sense of security, will be more likely to invest in the improvement of their homes and living environments than they would without legal tenure. However, the commodification of housing through the granting of land tenure has also led to changes in tenure structure and in the exclusion of the poorest households who cannot afford the market prices (Nientied et al 1982; Baross and van der Linden 1990). There are a number of tenure arrangements in operation in Amritsar's low-income housing system which reveal a number of findings. Legalisation has been gradually introduced in Amritsar's private self-help settlements, though the effects of upgrading have only had nominal effects. The survey shows that there is a relationship between physical structure and the possession of legal title (Table 8.4).

Table 8.4 Legal Ownership and Physical Housing Structure

| | Jhuggies | | Private Self-Help | | State-Assisted | |
	No Registry	Registry	No Registry	Registry	No Registry	Registry
Pacca	--	--	4	8	17	35
Semi-Pacca	--	--	31	45	81	65
Kaccha	56	--	64	47	2	--
Jhuggi	44	--	1	--	--	--
Total	100	0	60	40	68	32

Entries are column percentages.

All *jhuggi* settlements, by definition of the typology, are lacking legal tenure status (See Chapter Four). *Kaccha* and *jhuggi* structures are both prevalent material conditions of housing within *jhuggi* settlements. *Kaccha* structures are, however, representative of a heightened sense of security than are *jhuggi* structures. Bangla Basti and Righu Bridge are two *jhuggi* settlements which have a majority of *kaccha* structures due to their

161

ambiguous, and not outright illegal, status. Therefore, residents have been inclined to build their houses with semi-permanent, mud materials which, while requiring intensive labour inputs and upkeep, are relatively inexpensive.

Private self-help settlements have a diverse land tenure profile. Gujjarpura, Lahori Gate and Angarh have areas within each settlement which have been granted legal tenure with other areas which are either awaiting official notification or which are disputed with other agencies. Indira Colony has not been granted even partial legal tenure, though the Urban Basic Services Program (UBSP) has been introduced in the settlement giving residents a sense of security that the government will be less likely to evict them if their settlement is being installed with services. However, those households without registries are predominantly *kaccha* being 64 percent of all non-registry households while those houses with registries show a more even distribution between *kaccha* and *semi-pacca* structures, 46.7 and 45.3 percent, respectively.

State-assisted settlements show that a majority of households do not possess registry for their homes despite the fact that many have been given allotments. The inefficiency of bureaucratic procedures in ensuring the delivery of registries to official allotees is at least partially to blame for this. Only one-third of all state-assisted households have registries while two-thirds do not have registries. A majority of houses without registry have *semi-pacca* houses while a slightly less majority of all households with registries are also living in *semi-pacca* structures. The distinction between Municipal Corporation tenements and public housing schemes has been made in previous chapters which can be one of the attributes of these figures. Municipal Corporation tenements have not been awarded legal titles since most were built during the 1960's and 1970's as emergency housing. However, a relationship between legal tenure and building structure can be identified from the sample (See Table 8.4).

Private self-help settlements which have become legalised are Angarh, Lahori Gate and Gujjarpura. The methods of legalisation have varied throughout the settlements. Gujjarpura and Angarh are partition settlements which have in the past two decades been legalised through the granting of land titles. Gujjarpura has been upgraded through the installation of underground drainage pipes. Water supply has been left to individual households to invest in their own hand pumps. Angarh has been upgraded through the Sanitation Program administered through the Municipal Corporation. Latrines were installed in approximately fifty households in Angarh. Communal water pumps and toilets were also installed for the entire settlement. Residents of these two settlements, however, generally see the local government's actions towards partition settlements as far from adequate. This general attitude is illustrated in a statement by Balbir Singh, an Angarh resident:

> We do not even know who to talk to in the (Municipal) Corporation. None of us know if we have any rights to demand anything...all we know is that the land belongs to the Revenue (Waqf) Board. They have given some registries, but it took a long time to receive them and that was done through the help of political contacts.

A resident of Gujjarpura, Kulwant Singh, notes the effects of legalisation upon the settlement:

> My uncle got the registry for his house in 1973. He sold his house to our family and moved to another colony on the other side of the city. I don't know if he is better off there since he still lives in a *kaccha* house...but I know a lot of people who sold their houses once they received registries. I don't think we will sell our house (but) now that we have the registry at least we know that it is worth something.

The granting of tenure to residents in previously illegal or ambiguous settlements has in some cases led to the sale of houses, in the case of Kulwant Singh, sale to other family members. While the reasons for sale can be argued to be due to commodification processes, the low standards of living conditions in such settlements are also to do with the lack of adequate facilities and general low living standards.

The opportunity to leave unserviced settlements for better accommodation in other settlements is given to those households who have acquired registries by selling their homes for cash. This is a pattern which is gradually emerging, but which is not significant as yet to constitute a path within the low-income system due to the irregularity of regularisation schemes. However, with regard to access, legal land tenure has a direct impact upon the outward mobility of residents from formerly illegal settlements.

Another aspect of tenure relating to access is whether households own or rent their accommodation. While the survey reveals that renting does not comprise a considerable proportion of the overall housing typology, the status of migrant and non-migrant households within the three types of housing shows a higher representation of tenancy among migrant households (Table 8.5). Tenure patterns have been used in this study to identify the tenure arrangements under which social groups are accessing housing. *Jhuggi* settlements have been least affected by commodification processes with their outright illegal status and only a small representation of renting. The poorest migrant communities are drawn to *jhuggi* settlements due to their accessibility and low costs. The physical and legal distinctiveness of *jhuggi* settlements from other settlements also symbolises the marginality of the poorest communities within the city's social structure. The legalisation of private self-help settlements is the most dynamic sector of the low-income housing system which shows evidence of renting and resale.

Table 8.5 Tenure Status of Migrants

	Jhuggies		Private Self-Help		State-Assisted	
	Migrants	Non-Migrants	Migrants	Non-Migrants	Migrants	Non-Migrants
Own	89	--	93	95	62	89
Rent	11	--	7	5	38	11
Total	100	--	33	67	48	52

Entries are column percentages.

The social profile of private self-help settlements shows that they have the most diverse make-up. The state-assisted settlements, on the other hand, have a high proportion of renters, high caste communities and upper strata employment groups. These patterns reveal the effects upon social access that state-assisted self-help housing has had.

Broadly, the central finding made in this chapter is that the paths to housing access in *jhuggies*, private self-help and state-assisted settlements determine the social dimensions of access. The performance of the low-income delivery systems have exhibited an interception of state-assisted housing by higher income and social status groups. Private self-help settlements have a more varied social and tenure structure due to the prevalence of partition households and the subsequent settlement of local urban poor households. Pre-partition residents, partition migrant communities as well as legalised *jhuggi* households comprise the diversity within private self-help settlements. Meanwhile, new migrants have been occupying the worst housing, namely in *jhuggi* settlements. The new migrants coming from less developed parts of India have remained in peripheral *jhuggi* settlements from where the option to move is not accessible to them. Thus, as evident in the typology, social hierarchies and divisions have been solidified by state-initiated and state-supported schemes. In the case of state-supported programmes in private self-help settlements, legalisation has had the effects of raising the commercial values of housing and land and in increasing the incidence of renting. In state-initiated schemes the attainment of allotments by higher socio-economic groups has resulted in the further exclusion of the poorest, low-caste communities. Through both the neglect and attention by the relevant state authorities, access to housing has become socially differentiated.

---

[1] This is similar to the case of Dhaka where the city squatter settlements act as 'stepping stones' for new migrants who cannot afford the rents in inner-city slums (Shakur 1988).

[2] A common pattern of industrial intra-urban residential location is that the elites move outwards from the city centre and the poorest groups occupy the old city core.

[3] See Chapter Six, Table 6.6.

[4] For a more detailed discussion of the relationship between class analysis to caste in India see K.L. Sharma (1994) 'Caste and Class in India: Nexus, Continuity and Change' in *Social Stratification and Mobility*.

[5] Government workers have been categorised as anyone employed by the public sector. Therefore, many government workers living in private self-help settlements are sweepers which is a low level job while government workers in state-assisted households are predominantly low or middle ranking clerks or officers.

9   Conclusion

The growing crisis of poor households in Third World cities to access decent shelter and urban services is one which is likely to continue within the current trends of the "down-sizing development state." The detraction of resources away from direct assistance to partial assistance has far reaching implications and ones which will increasingly follow the market rather than basic needs. The manner in which the urban poor will struggle to survive in this context will continue to be of academic and political concern. This book is intended to provide neither closure on discussions of poverty nor a concrete formula for further policy towards housing the urban poor in the Third World. Instead, it has been intended to reflect upon the significance of social access in the experiences of the low-income housing sector as a means for offering scope for future studies on the political economy of housing. At best it is hoped that the study will incite both a critical understanding and new actions for change towards improved social equity.

The empirical side of this study has illustrated how the differentiating effects of self-help housing in its unaided and state-assisted forms have led to the inclusion of some social groups and the exclusion of others. The study has based its analysis upon social variables in the typology designed for the Amritsar context which exhibit a number of patterns of social differentiation. A central observation in the study is that the types of housing that are available in Amritsar's low-income housing system are hierarchical and that the social groups which are accessing each type of housing reflect this hierarchy in terms of their social composition and degrees of marginality.

The theoretical review of literature in the first chapter highlighted many of the central themes in approaches to poverty and housing in Third World cities. Commonly held constructions of legality and slum and squatter formation were critically outlined as often inadequate descriptions of the shelter situations of the urban poor, and the case for new ways in approaching shelter issues was presented through the unregulated housing sub-markets. The diversity of housing systems exhibited the distinctions between illegal and legal forms, settlement processes and housing structures challenging the previous tendencies in the literature to use broad

166

generalisations inappropriate to an evolving and changing sector. Rather, the case was made for an approach which acknowledges the non-exclusivity of the roles of the state and market. This culminated in a conceptual continuum rather than a simplistic dichotomous model of housing approaches, as outlined in Chapter Four, in forming the basis of the typology of *jhuggies,* private self-help settlements and state-assisted settlements.

The position of the poor in Amritsar is reflective of the marginal position that the city itself occupies in the South Asian and Punjab context. The history of the walled city beginning in the sixteenth century to the tearing down of the wall in 1921 by the British has played an integral role in the spatial and social layout of wider residential settlement in the city. The manner in which the highest concentration of low-income settlement has taken place in the area surrounding the walled city resembles that of other South Asian cities in which the old core of the city has had a magnetic effect in attracting the poor such as in Delhi, Lahore and Calcutta. However, the centrality of Amritsar within pre-colonial Punjab was undermined after the British accession of Punjab in 1849. Similarly, the high growth rate of cities such as Delhi and Karachi after partition due to in-migration differs from the Amritsar case in which the city suffered a decline following the out-migration of the resident Muslim community and the diversion of in-coming Hindu and Sikh refugees away from Amritsar towards other destinations in the newly demarcated India. The partition was the single most significant event affecting the poor as it divided communities along the lines of religion virtually erasing the once-majority Muslim community in Amritsar.

The post-colonial development of Amritsar within India has further sidelined the city as a border town. Therefore, the low-income settlements existing around the walled city in Amritsar are a symbol of the city's development and decline in their static and structurally marginal position. The city of Amritsar experienced a number of set-backs through the annexation of Punjab by the British empire in 1849, partition in 1947, the 1966 linguistic reorganisation of states and, most recently, the storming of the Golden Temple in 1984 and the situation that followed. Meanwhile, the green revolution-style agricultural development which has traditionally been associated with the region of Punjab has not acted as a catalyst for urban growth as it has for other Punjabi cities such as Ludhiana and Jullundur. This, as argued in Chapter Three, had profound effects upon the socio-economic prospects for the poor. The historical decline of the city caused changes to the labour force and hence decreasing employment opportunities available to them.

Opportunism by local residents and the disadvantage of migrants also was found to have left a mark on the ways in which communities have been formed and have consolidated. The historical phases of low-income settlement revealed that the residents of the walled city prior to partition in 1947 were able to occupy evacuee homes while refugees coming from West Punjab were left with the remaining Muslim evacuee houses or to build their own make-shift houses upon government land. This marked the establishment of squatting upon public land as a defined path to low-income settlement in the city.

More formal methods of housing development were also examined through public housing programmes both to upgrade the existing housing stock and to provide core housing. Government *ad hoc* programmes have remained insignificant in comparison to the self-help forms of housing provision due to the inability of the poorest groups to gain access to allotments and in the geographically distant location of such sites from centres of employment, therefore increasing travel costs to work. The low quality of housing and services in state-assisted houses further emphasised the inadequacy of state-level action where, for example in Chapter Seven, the housing conditions in state-assisted and self-help settlements in the survey did not differ significantly from one another in the survey.

The occupational activities of the households of the settlements around the walled city were identified, and one particular finding emerged that unskilled manual activities were the most common income generating activities in which poor households around the walled city are engaged. The absence of a large manufacturing sector and an increasingly obsolete crafts and skilled textiles sector has led to the overall decline in wages of such skilled professions in Amritsar while self-employed unskilled manual activities such as rickshaw pulling, construction, cleaning and painting have become the most accessible means of employment for the poor. The opportunities for upward occupational and income mobility were found to be limited.

In Chapter Seven the settlement processes of households around the walled city were examined through a detailed comparison of Amritsar and other studies of Third World cities. The varied nature of private self-help settlements with regard to settlement history, security of tenure and socio-economic make-up is cause for the diverse representation of housing types in the sample. Although *kaccha* structures are the most concentrated in private self-help settlements, a considerable proportion of *semi-pacca* houses were also found. In contrast, *jhuggi* settlements were entirely made

168

up of *jhuggi* and tent structures due to their illegal tenure status and reluctance of residents to invest while state-assisted settlements were comprised of predominantly *semi-pacca* and, to a lesser extent, *pacca* structures due to their legal status and in the prefabricated construction methods applied by the public housing sector. With regard to infrastructure the *ad hoc* upgrading methods of low-income settlements in Amritsar has mainly been done through the Urban Basic Services Programme (UBSP). As in the cited studies of Hyderabad, Pakistan (Siddiqui and Khan 1994) and of three Indian cities, Bhubaneshwar, Vijayawada and Delhi, (Bannerjee and Verma 1994), the potentials of infrastructural improvement of already existing settlements are restricted by housing and land supply mechanisms and in the bureaucracy's unwillingness to intervene.

The types of settlement in Amritsar were compared with those prevalent in other urban Third World contexts. In contrast to the organised mass invasions noted in Latin America (Gilbert and Ward 1982; Peattie 1982; Alexander 1988), it was found in Amritsar that a shift has taken place from mass invasions, which had occurred during the chaos of partition settlement and the several decades thereafter, to individual illegal occupation, similar to other South Asian cities such as Dhaka (Shakur 1988), Delhi (Mitra 1990) and Allahabad (Misra 1990). In the above mentioned South Asian cities the rapid commercialisation of housing and land markets limits the case for direct comparison as commercialisation has, due to the Galliara Scheme and increased market value of the land surrounding the walled city, only recently begun to alter the nature of settlement in Amritsar's low-income settlements. However, the slow rate of legalisation by public agencies has postponed the expulsion of the poorest households and in the entrance of middle-income groups as illustrated in Nientied's model (Chapter Seven, Figure 7.2). This model was presented as a prediction of possible trends in Amritsar once legalisation and commercialisation processes become more developed.

Chapter Eight combined the socio-economic and housing variables from the previous two chapters while also applying the theoretical arguments between the self-help and neo-marxist analyses of access to housing. The paths to housing in Amritsar's low-income settlements were illustrated through a schematic diagram incorporating the typology developed in the study (Figure 8.1). The entry points and routes which are available were analysed in terms of the social groups which are gaining access. The historical development of each path within the low-income housing system was traced, and the options available to newly arrived migrants were identified as being primarily within *jhuggi* settlements due to their limited economic power and relatively low social status. Other social groups such as partition migrants and local residents have each found their way within the housing system according to the time of settlement within Amritsar's history. The most significant paths to low-income settlement were established at the time of partition when Muslim evacuee homes were occupied by local residents of the walled city and by partition refugees coming from West Punjab. The practice of squatting on vacant government-owned land around the walled city at the time of partition was tolerated, and thereafter squatting around the walled city served the purpose of housing newly arrived migrants.

In opposition to Turner's model of migrant housing mobility, it was argued that the movement of different social groups within the city's housing system is structurally determined by both the supply of housing and facilities and the socially defined needs to housing (Castells 1977). Rather than attempting to form a generalised model of household development, the typology was utilised to outline the structural boundaries of the housing system within which the various groups are vying for access. The privileged status of state-assisted housing in Amritsar was clearly highlighted through the social make-up of residents in public schemes. Despite the aims of state housing to reach the poorest groups, the survey revealed a concentration of higher paid professions of skilled, professionals and government workers in state-assisted settlements. This exemplified a correlation between higher status jobs and the type of accommodation households occupy. Similarly, it was found that those households with higher income earnings were disproportionately represented in state-assisted housing. Even though state-assisted schemes require a small sum of initial cash payment and comparatively low monthly instalments to the private sector, the poorest have been excluded and remain in illegal *jhuggi* settlements while those with access to savings or capital are able to enter the housing market either through private self-help settlements or through official allotments. However, even where land and housing in *jhuggi* and private self-help settlements had once been provided to low-income households through non-commercial means such as squatting, the ability to obtain free land or housing in Amritsar is becoming more difficult due to the increasing commercial value of land close to the centre of the city.

This book has examined the social dimensions to housing access among the poor, and in doing so, the theoretical basis of the study has been positioned within the neo-marxist critique of self-help housing policies as a means of engaging in the   theoretical, methodological and ideological debates around the roles of the state, individual and market forces. Access to housing has come to be differentiated by a number of social and economic indicators of the household. While self-help has been found to be little more than people being left to house themselves, the role of the state in housing schemes targeted at low-income communities has shown to be a cause for further social differentiation in the housing market rather than as a provider of social welfare to the most needy of shelter. Thus, the neo-marxist analysis of state-supported self-help and of the state's collusion with capitalist modes of housing development have been upheld in this study, despite the orthodoxy of self-help within the literature.

The range of experiences of self-help housing in practice highlighted in this study have shown that it has neither been able to meet its own objectives of increasing supply to the demand nor has it been effective in achieving socially equitable access. Beyond the theoretical and ideological conflicts in the literature, the voice of one respondent (a partition migrant) in the study is perhaps the most pertinent voice to express the burden of self-help in an increasingly competitive housing market:

> As time has gone by, ...we thought that our conditions would have improved,...but they haven't. The harder we work, the less we seem to be able to afford.....

Appendix 1

Table A1.1 Percentage of People Living Below the Poverty Line in the States of India

India/States	1983-1984		1987-1988	
	Urban	Total	Urban	Total
All India	28.1	37.4	20.1	29.9
States				
Andhra Pradesh	29.5	36.4	26.1	31.7
Arunachal Pradesh	0.0	0.0	0.0	0.0
Assam	21.6	23.5	9.4	22.8
Bihar	37.0	49.5	30.0	40.8
Gujarat	17.3	24.3	12.9	18.4
Haryana	16.9	15.6	11.7	11.6
Himachal Pradesh	8.0	13.5	2.4	9.2
Jammu & Kashmir	15.8	16.3	8.4	13.9
Karnataka	29.2	35.0	24.2	32.1
Kerala	30.1	26.8	19.3	17.0
Madhya Pradesh	31.1	46.2	21.3	36.7
Maharashtra	23.3	34.9	17.0	29.2
Manipur	13.8	12.3	0.0	0.0
Meghalaya	4.0	28.0	0.0	0.0
Mizoram	0.0	0.0	0.0	0.0
Nagaland	0.0	0.0	0.0	0.0
Orissa	29.3	42.8	24.1	44.7
Punjab	**21.0**	**13.8**	**7.2**	**7.2**
Rajasthan	26.1	34.3	19.4	24.4
Sikkim	0.0	0.0	0.0	0.0
Tamil Nadu	30.9	39.6	20.5	32.8
Tripura	19.6	23.0	0.0	0.0
Uttar Pradesh	40.3	45.3	27.2	35.1
West Bengal	26.5	39.2	20.7	27.6

Source: Basic Statistics Relating to Indian Economy, Centre for Monitoring Indian Economy, Vol. 2, Sept. 1990 and 1991 in NIUA (1993*) Handbook of Urban Statistics.*

Appendix 2 Survey Settlements

Type 1: *Jhuggi* Settlements

Bangla Basti

Bangla Basti took its name from the regional origin of its residents, *Bangla* meaning Bengali and *basti* a common word used in Calcutta and other cities in South Asia for types of low-income settlements. Most people living there have migrated from outside Punjab, namely West Bengal and U.P. People have been settled in Bangla Basti since 1975 when a Congress Member of Legislative Assembly (MLA) had them forcibly relocated from outside Durgiana Mandir, a common first place of settlement for newcomers to Amritsar. At that time, the area was completely vacant. It was a barren site next to the *ganda nala* -moat- which was being used as a dumping ground. Being an undesirable place for other forms of settlement, it was an undisputed area of land, hence ideal for relocating these people. Around 60 to 70 jhuggies were uprooted from the Durgiana Mandir and nearly all of these people have ended up staying at Bangla Basti.

In 1986 and 1987 the Municipal Corporation tried to evict the residents from the land several times but were unsuccessful due to resistance from residents and apparent intervention by local politicians. Since then, the area has had no other threats of eviction or demolition. More recently, people have begun selling their houses and also renting their houses. With this sense of security, the commercial value of houses and land in Bangla Basti has risen ten times since 1986 where people bought houses and land for between Rs. 3000-4000. The value now for the same properties ranges from Rs. 30000-40000. However, with the Municipal Corporation's plans for sealing the *ganda nala* and building a storm water channel, it seems likely that the Municipal Corporation may attempt, perhaps more persistently this time, to evict the residents. The Divisional Town Planner's Office in Gwal Mandi had extensively studied the area surrounding the north west part of the outer circular road, where the highest concentration of low-income houses have been constructed and also where the storm water channel project is to be carried out in addition to other refurbishment activities. Preparations were being made in 1995 by the Municipal Corporation for the 'refurbishment' of the entire area by way of a shopping district to service high income groups. It is likely that these people will be forced to leave their homes and either entirely evicted or pushed back away from the outer ring road to land less proximate to the walled city.

From 1975 to 1980 there were no water sources in the settlement and residents walked 1-2 kilometres into the walled city in order to access water from public taps. In 1980 the residents built a *mandir* on the central road of the settlement. This was done for the purpose of ensuring security from further eviction. Residents had become aware of a legal clause which states that where there is a religious building neither the building itself nor any residential structures around it can be destroyed. Financial assistance for the construction of the *mandir* had been given by local political interests who were looking for votes in the upcoming elections. One tap was fitted to the outside of the mandir which became the sole source of water supply for the entire settlement. There are four public toilets which have been since constructed by the Municipal Corporation. They are pay facilities for which residents must pay one rupee and are not popularly used because of lack of upkeep and, most importantly, because there are no doors on the latrines. However, hand pump that is connected outside serves as a communal clothes-washing area as well as the second source of water for the area. Residents are outraged that the toilets were built in such an incomplete, haphazard manner. They feel that the expense of building them could have been better spent elsewhere.

The most striking feature of Bangla Basti that separates it from the other settlements studied is the amount of plastic bags, other plastic recyclable goods, and glass. Even when passing Bangla Basti from the outer circular road the smoke and plastic can be seen from a distance. As was evident in other particularly economically poor areas, recycling or scavenging work (*kabaardi*) is one of the most common sources of income. *Karaardi* is one of the least-paid jobs and is highly laborious. Scavengers roam the city in search of plastics for which they receive one rupee per kilogram of plastics collected. A majority of people in Bangla Basti do *kabaardi* work and therefore competition forces many to travel far from home. Local dealers buy the plastics and glass from them and then pass them on to recycling centres. Generally, it is women who handle the glass, and this is done with bare hands. Children also work in the recycling industry, though mostly in the scavenging side of the work (See Figure 6.4).

Durgiana Mandir

Durgiana Mandir is a Hindu temple which is a main religious site in Amritsar, second only to the Golden Temple in its attraction of tourists from other parts of India and abroad. Because of the continual presence of squatters around Durgiana Mandir for the past few decades, the 100 jhuggies which are located just outside of the temple grounds have come to be closely associated with the area around the temple. In this study the *jhuggi* settlement itself is referred to as Durgiana Mandir, though it is also known as Indira Nagar (See Figure 8.2). This site has been a popular first stop for newly arrived migrants. Current residents are primarily from West Bengal and have lived in the area since the early 1970s intermittently. The active role that the police have played on behalf of the temple administration in evicting the residents has created a history of eviction and resettlement. Because of the proximate location to the walled city, the market of tourists and the water and bathing facilities that are available on the temple grounds, even residents who have been evicted once or more have been persistent in staking their claim upon the site. A residents' association has been established with appointed officers, inferring an organised community. Many households which have been evicted and relocated to other areas such as Bangla Basti have chosen not to return due to the high level of insecurity. Many households living in Durgiana Mandir today are among the earliest settlers while others have come more recently. Durgiana Mandir has over the years provided immediate land to those who could not find accommodation elsewhere.

Mall Mandi

There are approximately 100 jhuggies on this plot of land which is located on the way out of Amritsar towards Jullundur on the Grand Trunk Road. The portion of land closer to G.T. Road is owned by the Municipal Corporation and is planned to be developed into a modern housing complex by the Punjab Housing and Development Board for middle income groups (MIG's) and high income groups (HIG's). The portion of the plot further back is owned by private owners. Because the land is low-lying and suffers from flooding, it is no longer useful as agricultural land. The private owners have had the land filled since every year the entire area is flooded during the rainy season, making it uninhabitable. Some of the raised portions of the land have been subdivided and sold to private interests while the area closer to the proposed housing development has been sold to the Punjab Housing and Development Board. The squatters here only have a sense of temporary security now that the land has been sold and that they are only waiting for when the development will begin. Residents are certain that they will inevitably be evicted and forced to find someplace else to live. Many residents have already been evicted numerous times from locations closer to the city: the pavement in front of Sangam cinema and Durgiana Mandir are two common places for squatting.

The colony is divided into two sections between the residents of Punjabi origin and those from eastern regions of India, namely U.P and Bihar. While there is a physical division between the two areas by means of a low brick wall which was actually built as the markers for the subdivisions that the owner has made for sale, the communities communicate with one another on only practical levels. There is a sense of cohesion among the Punjabis and among the U.P./Biharis. However, they all share a single water tap as the only source of water.

Many people only stay ten months each year and return to their homes in Bihar, U.P and rural parts of Punjab during the monsoon season. Most of the work done by the residents is toy making from shells. These are generally sold by the children around the city, particularly during festivals. Women do sewing/mending work for which they earn between ten and fifty paise per piece. Some embroidery work is also done by women for local cloth-selling shops.

Righu Bridge

Righu bridge is a major crossing over the railway lines. It also serves as one of the connections between the old and the new parts of the city. This community settled on this plot of land in 1947 as a planned invasion. As other communities were similarly scrambling for a place to live or to build their homes, this plot of land served a purpose to them due to the relative accessibility of water from the railway station connection taps. Righu Bridge is also close to a centre for employment, in this case, the railway station where shoe polishers traditionally flock. The residents living at Righu Bridge are a community of *mochis* or shoe polishers who make their living by soliciting the railway station for business.

Residents were continually harassed for many years and in 1979 they filed a case with the court for their right to stay to end the harassment. This particular plot of land lies directly across from a large Defense area and is officially owned by Defense. However, it was one individual high-ranking army officer who was laying claims to the land for the development of a transport business which eventually is what helped the Righu Bridge community to win their case in 1984. Since then, they have only been harassed once and have begun to build their homes with more solid materials. When they first settled, their homes were jhuggies of make-shift materials. Later, in the early 70s people began to construct their homes of kaccha materials. In 1984, with the winning of the case, most of the residents began to improve their homes with the sense of security from the injunction served to the army officer.

Type 2: Private Self-Help

Angarh

Angarh is divided in half by a single road. On one side, many people possess land titles while on the other side people do not possess land titles (See Figure 7.3). Those people who do possess titles are predominantly partition settlers while most non-partition settlers do not have titles. It was found that the entire area is under the ownership of the Waqf Board land and that prior to partition low-income groups which were predominantly Muslim were living here. This is significant in that Waqf Board land is donated for charitable causes and if low-income groups were given these donations of land to make their homes, then the contemporary 'tolerance' by the Municipal Corporation of people living on Waqf Board land is both a continuation and recognition of pre-partition, Islamic land law. When people came to settle there at partition, there were some built-up areas of predominantly low-income, basic housing structures with no services. Those who settled around 1947 occupied houses which were deserted by people who left to go to Pakistan. People who came afterwards occupied land which they built their homes upon. Eventually the area, including the empty land space eventually became consolidated up to the dumping ground where Indira Colony today exists.

In one area of the settlement, as part of the UBSP, several septic tanks were installed in individual houses with the opportunity to pay in instalments. The project has not been successful as some septic tanks have only been partially installed and others have not been properly installed. The shallow depth of the tanks has resulted in a severe problem of overflowing, which ultimately flows into the drains and has dangerous health implications. Particularly with private hand pumps being in use and no proper drainage system, the entire concept of installing safe, cost-effective sewerage has been defeated by incompetent structure and instalment. There are several public taps which are shared, but the number of private hand pumps is higher here than in other settlements. Because of its improved status since the early 1980s it has been the focus of Municipal Corporation improvement and upgrading efforts.

Residents seem to have a sense of security of not being evicted. It is a non-disputed area and the owner of the land, the Waqf Board, is the most sympathetic of public agencies towards the situation of the poor. Most people have the security that they will not be evicted as many people are now being given land titles through rental schemes with the Waqf Board. This is a sign that there are not plans to develop the land for any other purpose which could push them off the land. In 1965, around the time of the Indo-Pak war, residents were harassed by Defense. By the late 1980s, however, people were being given land titles. As a Congress stronghold, the settlement benefited from the local party activists' interest in obtaining votes from low-caste communities. Through this process of legalisation via political votes the commercialisation process was initiated. People began to sell their homes for profit with which they could afford to purchase a home with better facilities while others began renting rooms in their homes. Those who first settled at partition do not seem to have acquired registries while late comers have been more successful. This is an interesting settlement from the point of legality and land, as the area is divided in terms of legality, and in terms of sequential settlement: partition settlers, non- or post-partition settlers and commercial settlers. The relatively close vicinity to Bharariwal, a village, where many residents, predominantly women and children, work as agricultural labourers is a major source of income to the settlement.

Gujjarpura

This settlement is primarily a partition settlement. Before the time of partition the land had been used as both a grave sight for the local Muslim community and also as an open space for the British to keep and train their horses. At partition people who began to settle on the land were told that the land would not be disputed and were therefore allowed to stay. Being Waqf board land (particularly as a burial ground) the land has not been disputed and residents have not complained of harassment to vacate the land. There were two types of initial settlement:
1) people coming from Pakistan
2) people already living in Amritsar who saw the opportunity to acquire

land near the walled city to build their homes. The Municipal Corporation has attached some homes with a new sewarage system, which is blocked however and does not function. Some people have hand pumps but most people use the Municipal Corporation communal taps as the main source of water.

Despite the fact that most people do not have registries, there is a general sense of security due to the fact that there have not been threats by any local authority or private interests to demolish and take over the area. In 1960 the Waqf board auctioned off approximately 60 plots of land for Rs. 500-700 with registry. After this the area turned from a squatter colony to a slowly consolidating self-help area. The granting of registries to these few people meant that others without registries or potential settlers were given a speculative sense of security.

Indira Colony

The Emergency in 1975 caused the eviction of squatters around the city in other settlements. Squatter households came to this area to build houses for themselves. The area up until that time had been a dumping ground. In 1977 the new Janata Dal coalition government allowed squatters to stay, by and large, where they were. Subsequently, the colony received patronage from various political parties at a time when the struggle for political power was highly competitive and the vote bank of the poor became even more crucial. The widely contested attacks upon squatter communities during the Emergency resulted in promises made by the Congress Party to affected settlements for security of tenure hence it received its name after Indira Gandhi.

After 1980 many original squatters began selling their houses. Even though few people have legal ownership to the land, the settlement has received assurance from various political parties of its status. The UBSP scheme has been launched in the settlement to improve the water facilities. Due to the landfill upon which the settlement has been constructed, hand pumps do not provide safe enough drinking water and people have been forced to walk distances to access water. Only a few taps have been servicing the entire colony of 10,000 people which had been installed by politicians for votes. The Communist Party of India (Marxist) has also had extensive contact with local community leaders in Indira Colony in gaining political support through relief measures.

Many of the residents moved into the locality before or during partition, when houses were being deserted by people leaving for Pakistan. It was these deserted houses that became the homes of many people who were either homeless or who were eager to leave their congested living conditions within the walled city. The opportunity to move away from the city was made available to them. However, in this case the Waqf Board has allowed residents to stay on a monthly instalment scheme of 2200 rupees per month which implies the granting of a land title. This does not apply to the entire settlement as the houses near to Islamabad Road are acknowledged by the Waqf Board while houses further away and nearer to the are claimed by Defense and a local school have not been given titles and have rather been harassed by the Defense who threaten to demolish their homes if they do not leave.

There is evidence that some residents have purchased 'rights' to land from private squatters without any agreement for land title from between 1500-2000 Rupees per month. Some houses have private toilets as there is a sewerage connection in the area which is working in some streets but not in others. There are also pay toilets which were installed by the Municipal Corporation, though they are not maintained and people do not use them. Most people get water from Municipal Corporation taps on the streets. Few people have hand pumps as in other private self-help settlements such as Indira Colony or Angarh.

There were many houses outside Lahori Gate which existed at partition which had been owned by Muslims who had left for Pakistan. However, as the houses filled, others came and settled on the land nearby. It is most likely that the houses were occupied by those already living in the walled city or early partition refugee arrivals and that the land was occupied by the latecomers. The settlement outside of Lahori Gate became the nucleus of a larger settlement which attracted squatters and other people who could not afford housing in the walled city or elsewhere and who gained a certain amount of security by living alongside people who had been granted registries by the government. There was less likelihood that they would be evicted. Also the taps were installed by the Municipal Corporation for those who had been allotted land. Squatters could also benefit from this water source. More recently there has been a dispute over the land between the army and the Municipal corporation. The army has harassed residents without registry while the municipal Corporation has not been active in trying to recover the land.

Type 3: State-Assisted

Ajnala Road

The colony was constructed in 1982 and services high, middle and low-income groups. There have been several waves of settlement that have occurred. First, the original allotees who settled have received registries and then chosen to live in the flats or after receiving registries have either rented or sold the flats. The basic cost of the flats is Rs. 25,000 and the market price in 1995 varied between 30000-50000 and 100,000 rupees. There are many households with above average income levels who are employed in secure, government jobs such as in the post office and customs. These residents have invested into their property by building extensions which has resulted in the increased valuation of many flats. Because of the openness of the space and the colony's relative closeness to Lawrence Road and the other commercial areas of north Amritsar, it is a desirable residential location for many households. However, the walled city is still the place where most low-income households rely upon income generation from where it is distant.

The flats of Ajnala Road are two-storey buildings with two rooms, kitchen and attached bathroom. The water pressure in LIG flats is so low that first floor residents do not get water and have to access water from ground floor neighbours or taps. The sanitation system is blocked such that in some areas of the colony it does not work and people are forced to use the open air. Residents are basically not happy with the services, but because of relative conditions elsewhere in the city as well as the location of the colony, most people have opted to stay.

Gwal Mandi

This settlement has been developed by the Amritsar Improvement Trust, though this includes an older set of flats which had been built by the Punjab Housing Board . There are seventy-two three-storey flats for low-income groups. Gwal Mandi reflects the contemporary arrangements that exist among different agencies in low-income housing development. The Municipal Corporation and the Improvement Trust play different, though collaborative, roles. The Municipal Corporation, who owns much land in Amritsar, has allocated plots for commercial and residential development to the Improvement Trust who further accepts tenders from private developers. The flats have been allocated to people who have applied for allotments through the Improvement Trust.

The settlement was first developed in 1972 when the Punjab Housing Board erected flats for low-income groups. However, the recently constructed flats in 1995 were at the time of the field survey just completed. The flats are basic and are small in size in comparison to the Ajnala Road flats with only one room. The similar problem of water pressure to first and second floor residents has made many residents dissatisfied with the accommodation, though the relative access to walled city, the spacious grounds around the complex and the security of tenure are positive aspects of the settlement.

Outside Gilwali Gate

These sixty flats were built in 1977 for the purpose of providing accommodation for sweepers working for the Municipal Corporation. However, there are very few families living in the flats who actually work for the Municipal Corporation. As soon as the flats were built non-allottees occupied them. Thereafter, the Municipal Corporation asked them to leave, but as the conditions of the complex have deteriorated no one has since bothered them. It appears that there were two waves of settlement. One occurred when people occupied the units as soon as they were finished. The second phase was when people occupied them once original occupiers had found other means of housing and sold them for nominal prices. Some relatives and friends of residents hearing of original occupiers leaving settled in the deserted flats.

The complex was built with services not fully installed. Residents have made their own open drain system from bricks and mud.. The Municipal Corporation had installed taps but there is no water connection. The residents all access water from public taps on the main road.

Janta Flats

This settlement, also known as known as Ranjit Avenue Improvement Trust flats, is a colony of 263 flats which were built in 1971 on land owned by the Municipal Corporation. There are a combination of single and multi-storey flats in the settlement.

Many people have bought the flats from original allotees, who made a profit from selling the flats. The market value of flats in 1995 ranged from 20,000 to 1.5-3 lakh rupees. From the sample taken from this settlement it is evident that predominantly high caste, middle-income groups are occupying the flats, and that the resale of flats is leading to the further exclusion of the poorest groups.

Roop Nagar

This colony of semi-detached housing was built by the Punjab Housing Board and then allotted to defense and government employees. There is strong evidence that the Housing Board was commissioned by defense and the local government to construct the colony for some of its middle ranking workers. The original allottees purchased the flats for 24,000 and 40,000 rupees, paying 365 rupees per month. The colony was not installed with privately attached water and sanitation provision. Due to the prime location of the colony to the walled city, many low-income residents have sold their houses for double and triple the original price and have moved to other public schemes with the large cash sums gained through resale. However, the houses are now occupied by predominantly middle-income households who are economically far better off than low-income households in nearby self-help settlements in the sample as they have higher steady incomes and employment security. Individual households have improved their houses which has contributed to the further increasing property values in the settlement.

Valmiki Quarters

The Valmiki Quarters are a set of 100 flats which were specifically built for scheduled castes. One portion of the flats were built in 1952 and were occupied by government sweepers, mainly from Amritsar. In 1995 a second portion of the flats were being constructed but do not figure in this survey.

   The facilities in the older flats are basic with private toilets, though sewerage is continually blocked, shared taps i.e. people washing clothes and dishes in the gullies, *kaccha* drains, and bricked pathways. Many of those people living in these flats are second generation. They are the children of original allottees who had been or still are Municipal Corporation employees. The children of the original allottees are not predominantly Municipal Corporation employees with these jobs becoming increasingly scarce. However, a new set of flats have been constructed in the last five years for current Municipal Corporation employees. The facilities are only slightly better because of their relative newness.

Waryam Singh Colony

This colony was built by the Municipal Corporation in 1965. It was built for the purpose of rehousing rural flood victims from Amritsar district and displaced people from the Indo-Pak war living near the border. No water connections or toilet facilities were ever installed, apart from two Municipal Corporation taps on the main road which are shared by all households. While a few flood and war victim families are living in the flats, in 1965 many other squatter households living in the area had heard of the allocation of flats by the Municipal Corporation and subsequently occupied them. There were no bureaucratic checks on who actually occupied the flats and therefore many flood and war victims were forced to find housing elsewhere.

Appendix 3

Table A3.1 Employment Categories used in the Census of India

Main workers - those who work for the major part of the year (at least 6 months).
Marginal workers - to include women/children involved in unpaid work (seasonal or irregular).
Cultivators - employer, single worker or family worker (working on someone else's land does not count).
Agricultural labourers - person who works on another person's land for wages or kind. No right of lease or contract on land on which works.
Household industry - participation of 1 or more members of household in production. Professional services conducted in house not included.
Other workers - all who are engaged in some type of economic activity during the year but who are not cultivators/agricultural workers are categorised as other workers.
Source: Government of India *(1992) Census of India 1991, Final Population Totals: Brief Analysis of Primary Census Abstract.*

Table A3.2 Employment Categories used in the British Census

	Major SOC Group*	Examples of Occupations
1	Managers & administrators	manager in local govt, health service administrator
2	Professional	doctor, architect, librarian, teacher
3	Associate professional & technical	nurse, lab technician, building inspector
4	Clerical & secretarial	secretary, clerk, cashier
5	Craft & related	sewing machinist, bricklayer, electronic production fitter, lathe operator
6	Personal & protective services	police constable, chef, hairdresser, nursery nurse, nursing auxiliary
7	Sales	sales representative, check-out operator
8	Plant & machine operatives	packer, electronic line worker/assembler, bus driver
9	Other occupations	agricultural worker, labourer

Source: Office of Population Censuses and Surveys (1990), *Standard Occupational Classification,* Volumes 1-3.

Bibliography

Abrams, C. (1964) *Housing in the Modern World: Man's Struggle for Shelter in an Urbanising World* Faber: London.

Abrams, C. (1966) *Squatter Settlement: The Problem and the Opportunity*, Department of Housing and Urban Development: Washington D.C.

Abu-Lughod, J. (1976) 'Developments in North African Urbanism: The Process of Decolonization' in B.J.L. Berry (ed.) *Urbanization and Counterurbanization* Sage: Beverly Hills, pp. 191-211.

Ahmed, V. (1986) 'Lahore Walled City Upgrading Project' in G. Shabbir Cheema (ed.) *Reaching the Urban Poor: Project Implementation in Developing Countries*, Westview Press: London.

Ahsan, F. (1986) 'Provision of Urban Services to the Urban Poor: A Case Study of Lahore Katchi Abadis' in G. Shabbir Cheema (ed.) *Reaching the Urban Poor: Project Implementation in Developing Countries*, Westview Press: London.

Aina, T.A. (1989) 'Africa's Current Crisis and the Urban Condition' *African Urban Quarterly.*

Aina, T.A. (1990) 'The Politics of Sustainable Urban Development' in Cadman, David and Payne, Geoffrey (1990) (eds.), *The Living City: Towards a Sustainable Future.*

Alavi, H. (1980) 'India: Transition from Feudalism to Colonial Capitalism' *Journal of Contemporary Asia*, Vol. 19, 359-399.

Aldrich, B. and Sandhu, R.S. (1995) (eds.) *Housing the Urban Poor: Policy and Practice*, Zed Books: New Delhi.

Alexander, E.R. (1988) 'Informal Settlement in Latin America and Its Policy Implications' in Carl Patton (ed.) *Spontaneous Shelter: International Perspectives and Prospects*, Temple University Press: Philadelphia.

Alvi, I. (1997) *The Informal Sector in Urban Economy: Low Income Housing in Lahore*, Oxford University Press: Karachi.

Amin, S. (1976) *Unequal Development: An Essay on the Social Formations of Peripheral Capitalism*, Harvester Press: New York.

Amis, P. (1984) 'Squatters or Tenants: The Commercialization of Unauthorized Housing in Nairobi' *World Development*, Vol. 12, no. 1.

Amis, P. (1988) 'Commercialized Rental Housing in Nairobi, Kenya' in C.V. Patton (ed.) *Spontaneous Shelter: International Perspectives and Prospects*, Temple University Press: Philadelphia.

Amis, P. (1990) 'Key Themes in Contemporary African Urbanisation' in Amis and Lloyd (eds.) *Housing Africa's Poor*, Manchester University Press: Manchester.

Amis, P. (1996) 'Long-run Trends in Nairobi's Informal Housing Market' *Third World Planning Review*, Vol. 18, No. 3, August, pp. 271-286.

Angel, S. (1983) 'Upgrading Slum Infrastructure: Divergent Objectives in Search of a Consensus' *Third World Planning Review*, Vol. 12, No. 1.

Angel, S., Archer, R.W., Tanphipat, S. and Wegelin, E.A. (1983) (eds.), *Land for Housing the Poor*, Select Books: Singapore.

Angel, S. and Ponchokchai, S. (1990) 'The Informal Land Subdivision Market in Bangkok' in Baross, P. and van der Linden, J. (eds.) *The Transformation of Land Supply Systems in Third World Cities*, Avebury: Aldershot.

Ansari, J. (1995) 'Towards a holistic approach for housing the poor in India' *Institute of Town Planners (ITPI) Journal*, Vol. 13, No. 3 & 4 (161 &162), pp. 39-53.

Armstrong, W. and McGee, T.G. (1983) *Theatres of Accumulation: Studies in Asia and Latin American Urbanization*, Methuen: London.

Asad, T. (ed.) (1973) *Anthropology and the Colonial Encounter*, Ithaca Press: London.

Asian Development Bank (1987) *Urban Policy Issues: Regional Seminar on Major Urban Policy Issues*, Proceedings from conference in Manila, February 1-3 1987.

Audefroy, J. (1994) 'Eviction trends worldwide- and the role of local authorities in implementing the right to housing' *Environment and Urbanization*, Vol. 6, No. 1, April, pp. 8-24.

Baken, R.J. and Rao, M. U. (1995) *The Bhaskara Rao Peta Connection: Low-Income Housing and Relocation in Vijayawada, India: The Views of Slum Dwellers and Slum Leaders*, Urban Research Working Papers, 38, Department of Cultural Anthropology/Sociology of Development, Vrije Universiteit: Amsterdam.

Ballard, R. (1987) 'The Political Economy of Migration: Britain, Pakistan and the Middle East' in J. Eades (ed.) *Migration, Labour and the Social Order*, ASA Monographs, No. 25, Tavistock: London.

Banga, I. (1996) 'Arya Samaj and Punjabi Identity' in Singh and Thandi (eds.) *Globalisation and the Region*, Association for Punjab Studies: Coventry.

Bannerjee, B. and Verma, G.D. (1994) 'Three Indian Cases of Upgradable Plots' in *Third World Planning Review*, 16 (3), pp. 263-276.

Bapat, M. (1981) *Shanty Town and City: The Case of Poona*, Pergamon: Oxford.

Baran, P. (1957) *The Political Economy of Growth*, New York Monthly Press: New York.

Baross, P. (1983) 'The Articulation of Land Supply for Popular Settlements in Third World Cities' in Angel et al (eds.) *Land for Housing the Poor*, pp. 180-210.

Baross, P. (1984) 'Kampong Improvement or Development? An Appraisal of the Low-Income Settlement Upgrading Policy in Indonesia' in Eugene Bruno et al, *Development of Low-Income Neighbourhoods in the Third World*, Archimed Publishers, Darmstadt.

Baross, P. and van der Linden, J. (1990) (eds.) *The Transformation of Land Supply*

Systems in Third World Cities, Avebury: Aldershot.

Batra, B.R. (1985) 'Urban Development in Punjab - Problems and Prospects in Terms of Housing, Transportation and Other Infrastructure' in S.N. Misra (ed.) *Urbanisation and Urban Development in Punjab*, Guru Nanak Dev University: Amritsar.

Beall, J. (1995) 'Social Security and Social Networks Among the Urban Poor in Pakistan' *Habitat International*, Vol. 19, No. 4, pp. 427-445.

Bhattacharya, K.P. (1990) 'Housing in India- observations on the government's intervention policies' in Gil Shidlo (ed.) *Housing Policy in Developing Countries,* Routledge: London, pp. 67-103.

Bradnock, R. (1984) *Urbanisation in India*, John Murray Publishers: London.

Brewer, J. and Hunter, A. (1989) *Multi-Method Research: A Synthesis of Styles*, Sage: London.

Bruno, E., Korte, A. and Mathey, K. (1984) (eds.) *Development of Urban Low-Income Neighbourhoods in the Third World*, Archimed Publishers: Darmstadt.

Bulmer, M. (ed.) (1977) *Sociological Research Methods: An Introduction*, Macmillan: London.

Bulmer, M. and Warwick, D.P. (eds.) (1993) *Social Research in Developing Countries: Surveys and Censuses in the Third World*, UCL Press: London.

Burgess, E.W. (1925) 'The Growth of the City: An Introduction of a Research Project' in R.E. Park et al (eds.) *The City*, University of Chicago Press: Chicago.

Burgess, R. (1978) 'Petty Commodity Housing or Dweller Control? A Critique of John Turner's View on Housing Policy' *World Development*, Vol. 6, Nos. 9-10.

Burgess, R. (1982) 'Self-Help Advocacy: A Curious Form of Radicalism. A Critique of the Work of J.F.C. Turner' in P. Ward (ed.) *Self-Help Housing: A Critique*, Mansell Publishing Ltd.: London.

Burgess, R. (1984) 'The Limits of Self-Help Housing Programs' in E. Bruno et al (eds.) *Development of Urban Low-Income Neighbourhoods in the Third World*, Archimed Publishers, Darmstadt, pp. 15-60.

Burgess, R. (1985) 'Problems in the Classification of Low-Income Neighbourhoods in Latin America' *Third World Planning Review*, Vol. 7, No. 4, pp. 287-306.

Burgess, R. (1992) 'Helping Some to Help Themselves' in Mathey, K. (ed.) *Beyond Self-Help Housing*, Mansell: New York.

Burgess, R.G. (1982) (ed.) *Field Research: A Sourcebook and Field Manual*, George Allen and Unwin Publishers Ltd.: London.

Byres, T.J. (1981) 'The New Technology, Class Formation and Class Action in the Indian Countryside' *Journal of Peasant Studies*, Vol. 8, No. 4.

Byres, T.J. (1983) *The Green Revolution in India*, Open University Third World Studies, Case Study 5, Open University Press.

Byres, T.J. (1994) *State Development Planning in India*, Oxford University Press: Oxford.

Cadman, D. and Payne, G. (1990) (eds.), *The Living City: Towards a Sustainable Future*, Routledge: London.

Castells, M. (1977) *The Urban Question: A Marxist Approach*, Edward Arnold Publishers: London.

Castells, M. (1978) *City, Class and Power*, The Macmillan Press Ltd.: London.

Castells, M. (1982) 'Squatters and Politics in Chile, Peru and Mexico' in Helen Safe (ed.) *Towards a Political Economy of Urbanization in Third World Countries*, Oxford University Press: Oxford.

Castells, M. (1983) *The City and the Grassroots: A Cross-Cultural Theory of Urban Social Movements*, Edward Arnold Publishers Ltd.: London.

Census of India 1991 Final Population Totals: Brief Analysis of Primary Census Abstract, Series 1, Paper 2 of 1992.

Chana, T.S. (1984) 'Nairobi: Dandora and Other Projects' in Geoffrey Payne (ed.) *Low Income Housing in the Developing World*, Wiley: London.

Cheema, G.S.(1986) (ed.) *Reaching the Urban Poor: Project Implementation in Developing Countries*, Westview Press: London.

Christopher, P. (1986) 'Planning and Development for Slum Areas in Ludhiana: A Case Study of Kirpal Nagar' an unpublished Master's dissertation submitted to Guru Nanak Dev University: Amritsar.

Collier, D. (1976) *Squatters and Oligarchs*, Johns Hopkins University Press: Baltimore.

Connell, J.B. et al (1976) *Migration from Rural Areas: The Evidence from Rural Studies*, Oxford University Press: Delhi.

Connell, J.B. et al (1976) *Migration from Rural Areas: The Evidence from Rural Studies*, Oxford University Press: Delhi.

Crook, N. (1993) *India's Industrial Cities: Essays in Economy and Demography*, Oxford University Press: Delhi.

de Witt, J. (1992) *The Socio-Poolitical Impact of Slum Upgrading in Madras: The Case of Anna Nagar 1982-1990*, Urban Research Working Paper 29, Institute of Cultural Anthropology/Sociology of Development, Vrije Universiteit: Amsterdam.

Desai, A.R. and Pillai, D.S. (1991) (eds.), *Slums and Urbanization*, Sangam Books Ltd.: London.

Desai, V. (1995) *Community Participation and Slum Housing: A Study of Bombay*, Sage: New Delhi.

Detereux, S. and Hoddinott, J. (eds.) (1992) *Fieldwork in Developing Countries*, Harvester Wheatsheaf: London.

Dickens, P. (1990) *Urban Sociology: Society, Locality and Human Nature*, Harvester Wheatsheaf: New York.

Dickenson, J.P. et al (1983) *A Geography of the Third World*, Methuen: London.

Dupont, V. (1995) *Decentralized Industrialization and Urban Dynamics: The Case of Jetpur in West India*, Sage: New Delhi.

Dwyer, D. (1974) *People and Housing in Third World Cities*, Longman: London.

Fernandes, K. (1994) 'Katchi Abadis Living on the Edge' *Environment and Urbanization*, Vol. 6, No. 1, April, pp. 50-58.

Franceys, R. and Cotton, A. (1989) 'Benefits and Sustainability in Infrastructure Provision: India and Sri Lanka' a paper presented at the Sixth Inter-Schools Conference on Development, 18-19 March at the University of Sheffield.

Frank, A. G. (1967) *Capitalism and Underdevelopment in Latin America*, Monthly Review Press: New York.

Frank, A. G. (1969) *Capitalism and Underdevelopment in Latin America*, Monthly Review Press: New York.

Frankel, F. (1971) *India's Green Revolution: Economic Gains and Political Costs*, Princeton University Press: Princeton NJ.

Garg, S.K. (1990) 'Reaching Low-Income Families: A Private Sector Approach' in Baross and van der Linden (eds.).

Gauba, A. (1988) *Amritsar: A Study in Urban History (1840-1947)*, ABS Publications, Jullundur.

Gilbert, A. (1992) *Cities, Poverty, and Development : Urbanization in the Third World*, Oxford University Press: Oxford.

Gilbert, A. and Gugler, J. (1982) *Cities, Poverty and Development: Urbanization in the Third World*, Oxford University Press: Oxford.

Gilbert, A. and Ward, P. (1982) 'Residential Movement among the Poor: The Constraints of Housing Choice in Latin American Cities' *Transactions of the Institute of British Geographers*, New Series, Vol. 7, No. 2, pp. 129-149.

Gilbert, A. and Ward, P. (1985) *Housing, the State and the Poor: Policy and Practice in Three Latin American Cities*, Cambridge University Press: Cambridge.

Gill, R. (1991) *Social Change in Urban Periphery*, Allied Publishers Ltd.: Delhi.

Gill, S.S. (1996) 'Development Processes and the Problem of Punjabi Identity in Indian Punjab' in Singh and Thandi (eds.) *Globalisation and the Region*, Association of Punjab Studies: Coventry.

Government of India Registrar General (1981) *Census of India*, Controller of Publications: Delhi.

Government of Punjab (1994-95) *Economic Survey of Punjab*, Economic Adviser to Government Punjab: Chandigarh.

Government of Punjab (1995) Department of Housing and Urban Development, *Urban Development Strategy for Punjab*, Department of Housing and Development: Chandigarh.

Government of Punjab (no date) *Statistical Abstract of Punjab* 1947-50.

Government of Punjab (India) (1980) *Key Results of the Survey: Identification of Weaker Sections in Punjab*, The Economic Advisor to Government Punjab, Publication no. 362, Sept-Dec. 1980: Chandigarh.

Government of Punjab (1982) *Statistical Abstract of Punjab 1981*, Chandigarh.

Government of Punjab (1992) 'Management Model and Organisation Structure for Alleviation of Poverty for the Urban Poor in Punjab,' Town and Country Planning and Local Self-Government Departments: Chandigarh.

Grewal, J.S. and Banga, I. (1979) (eds.) *Studies in Urban History*, Guru Nanak
190

Dev University Press: Amritsar.

Grimes, O.F. (1976) *Housing for Low-Income Urban Families: Economics and Policies in the Developing World*, World Bank, Johns Hopkins University Press: London.

Gugler, J. (1988) (ed.) *The Urbanisation of the Third World*, Oxford University Press: Oxford.

Gugler, J. (1982) 'Overurbanization Reconsidered' *Economic Development and Cultural Change*, Vol. 30, 173-189.

Gupta, D. (1985) *Urban Housing in India*, World Bank Working Paper, No. 730, Washington D.C.

Gupta, D.G., Kaul, S. and Pandey, R. (1993) *Housing and India's Urban Poor*, Har-Anand: New Delhi.

Hakim, C. (1987) *Research Design*, Routledge: London.

Hamberg, J. (1990) 'Cuba' in Mathey, Kosta (1990) (ed.) *Housing Policies in the Socialist Third World*, Mansell: London.

Hamza, M.M. (1994) 'Community Participation: Barriers to Effective Community Participation in Urban Upgrading in the Third World' unpublished paper presented at the 11th Inter-Schools Conference on Architecture and Planning in the Developing World, University of York, March 28-29, 1994.

Hardoy, J. E. (1983) 'The Inhabitants of Historical Centres: Who is Concerned about their Plight?' *Habitat International*, Vol. 7(5/6), pp. 151-162.

Hardoy, J.E. and Satterthwaite, D. (1989) *Squatter Citizen*, Earthscan: London.

Hardoy, J. E., Satterthwaite, D. and Cairncross, S. (1990) (eds.), *The Poor Die Young: Housing and Health in Third World Cities*, Earthscan: London.

Harms, H. (1982) 'Historical Perspectives on the Practices and Purpose of Self Help Housing' in Ward (1982) (ed.) *Self Help Housing: A Critique*, Mansell: London.

Harms, H. (1992) 'Self-Help Housing in Developed and Third World Countries' in Mathey (1992), *Beyond Self-Help Housing*, Mansell: London.

Harvey, D. (1985) *The Urbanization of Capital: Studies in the History and Theory of Capitalist Urbanization*, Johns Hopkins University Press: Maryland.

Harvey, D. (1989) *The Urban Experience*, Basil Blackwell: Oxford.

Hasan, A. (1987) *A Study of Metropolitan Fringe Developments in Karachi; Focusing on Informal Land Subdivision*, Mimeo, UNESCAP: Karachi.

International Labour Organisation (ILO) (1972) *Employment, Incomes and Equity: A Strategy for Increasing Productive Employment in Kenya*, ILO: Geneva.

Jayaram, N. and Sandhu, R.S. (1988) (eds.) *Housing in India*, B.R. Publishing Corporation: Delhi.

Jeffrey, R.(1986) *What's Happening to India?: Punjab, Ethnic Conflict, Mrs. Gandhi's Death and the Test for Federalism*, Macmillan: London.

Jha, S.S. (1986) *Structure of Urban Poverty: The Case of Bombay Slums*, Popular Prakashan: Bombay.

Jodha, N. (1988) 'The Poverty Debate in India: A Minority Viewpoint' in *Economic and Political Weekly*, Vol.23, No 2421-2428.

Johar, R.S. and Khanna, J.S. (1983) (eds.) *Studies in Punjab Economy*, Punjab School of Economics, Guru Nanak Dev University Press: Amritsar.

Kemeny, J. (1992) *Housing and Social Theory*, Routledge: London.

King, A. (1975) *The Colonial City*, Routledge: London.

King, A. (1976) *Colonial Urban Development: Culture, Social Power and Environment*, Routledge and Kegan Paul: London.

King, A. (1990) *Urbanism, Colonialism, and the World-Economy: Cultural and Spatial Foundations of the World Urban System*, Routledge: London.

Kirke, J. (1984) 'The Provision of Infrastructure and Utility Services' in G. Payne (ed.) *Low-Income Housing in the Developing World*, Wiley: London.

Khan, S.A. (1987) 'Housing for the Urban Poor in Karachi' unpublished master's dissertation, University of Hawaii.

Khan, S.A. (1994) 'Attributes of Informal Settlements Affecting Their Vulnerability to Eviction: A Study of Bangkok' *Environment and Urbanization*, Vol. 6, No. 1, pp. 25-39.

Kohli, A. (1990) *Democracy and Discontent: India's Growing Crisis of Governability*, Cambridge University Press: Cambridge.

Kothari, M. (1994) 'Tijuca Lagoon: Evictions and Human Rights in Rio de Janeiro' *Environment and Urbanization*, Vol. 6, No. 1, pp. 63-73.

Krishan, G. (1996) 'The Integrated Development of Small and Medium Towns in Punjab' in Singh, K., Steinberg, F. and von Einsiedel, N. (eds.) *Integrated Urban Infrastructure Development in Asia*, Intermediate Technology Publications: London.

Kudaisya, G. (1995) 'The Demographic Upheaval of Partition: Refugees and Agricultural Settlement in India 1947-67' *South Asia*, 18, special issue, pp. 73-94.

Kumar, S. (1989) 'How Poorer Groups Find Accommodation in Third World Cities' *Environment and Urbanization*, Vol. 1, No. 2, October.

Kundu, A. (1991) 'Micro Environment in Urban Planning: Access of Poor to Water Supply and Sanitation' *Economic and Political Weekly*, Vol. XXVI, No. 37, September 14, 1991.

Laquain, A.A. (1979) 'Squatters and Slum Dwellers' in S.H.K. Yeh and A.A. Laquain (eds.) *Housing Asia's Millions*, International Development Research Centre: Ottawa.

Lea, J.P. (1979) 'Self-Help and Autonomy in Housing: Theoretical Critics and Empirical Investigation' in Murison and Lea (eds.) *Housing in Third World Countries: Perspectives on Policy and Practice*: London.

Lea, J.P. (1983) 'Customary Land Tenure and Urban Housing Land: Partnership and Participation in Developing Societies' in Angel, Shlomo, Archer, R.W., Tanphipat, S., and Wegelin, E.A. (1983) (eds.), *Land for Housing the Poor*, Select Books, Singapore.

Lewis, W.A. (1954) 'Economic Development with Unlimited Supplies of Labour' in *The Manchester School of Economic and Social Studies*, May, Vol. 22, pp. 139-191.

Lewis, O. (1966) *La Vida: A Puerto Rican Family in the Culture of Poverty*,

Panther: New York.

Lipton, M. (1977) *Why Poor People Stay Poor: Urban Bias in World Development*, Maurice Temple Smith: London.

Lipton, M. (1983) 'Poverty, Under-Nutrition and Hunger World Bank Staff' Working Papers, No. 597, World Bank: Washington.

Linn, J.F. (1983) *Cities in the Developing World: Policies for Their Equitable and Efficient Growth*, Oxford University Press: Oxford.

Lloyd, P. (1979) *Slums of hope? : Shanty towns of the Third World*, Penguin Books: Harmondsworth.

Lloyd, P. (1980) *The 'Young Towns' of Lima- Aspects of Urbanization in Peru*, London.

Luthra, K.L. (1949) *Impact of Partition on Industries in Border Districts of East Punjab*, Ludhiana.

Majumdar, P.S. and Majumdar, I. (1978) *Rural Migrants in an Urban Setting*, Transaction Books: New Brunswick.

Mangin, W. (1967) 'Latin American Squatter Settlements: A Problem and a Solution' *Latin American Research Review*, Vol. 2.

Mangin, W. (1970) *Peasants in Cities: Readings in the Anthropology of Urbanization*, Houghton Mifflin: Boston.

Marcuse, P. (1992) 'Why Conventional Self-Help Projects Won't Work' in Mathey, K. (ed.) *Beyond Self-Help Housing*, Mansell: London.

Marcussen, L. (1990) *Third World Housing in Social and Spatial Development*, Avebury: London.

Marsh, C. (1982) *The Survey Method: The Contribution of Surveys to Sociological Explanation*, George Allen & Unwin: Boston.

Marsh, C. (1988) *Exploring Data: An Introduction to Data Analysis for Social Scientists*, Polity Press: Cambridge.

Mascarenhas-Keyes, S. (1987) 'The Native Anthropologist: Constraints and Strategies in Research' in Anthony Jackson (ed.) *Anthropology at Home*, ASA Monographs 25, Tavistock Publications: London.

Mathey, K. (1990) (ed.) *Housing Policies in the Socialist Third World*, Mansell: London.

Mathey, K. (1992) (ed.), *Beyond Self-Help Housing*, Mansell: London.

Mavi, H.S. and Tiwana, D.S. (1993) *Geography of Punjab*, National Book Trust, India: New Delhi.

Mayer, A.C. (1980) *Caste and Kinship in Central India: A Village and its Region*, Routledge and Kegan Paul: London.

McAuslan, P. (1985) *Urban Land and Shelter for the Poor*, Earthscan: London.

Mehta, D. and Mehta, M. (1989) *Metropolitan Housing Market: A Study of Ahmedabad*, Sage: New Delhi.

Mehta, M. (1988) 'Urban Housing Policies: An Appraisal' in Jayaram and Sandhu (eds.) *Housing in India: Problem, Policy and Perspectives*, B.R. Publishers: Delhi.

Mehta, D. and Mehta, M.(1994) 'Affordable Housing and Urban Planning in India: An Overview of Issues and Policies' in R.N. Sharma (ed.) *Indo-Swedish*

Perspectives on Affordable Housing, Tata Institute of Social Sciences: Bombay.

Miah, Md. A.Q., Weber, K. and Islam, N. (1988) *Upgrading a Slum Settlement in Dhaka*, Division of Human Settlements Development, Asian Institute of Technology: Bangkok.

Mikkelson, B. (1995) *Methods for Development Work and Research*, Sage Publications: London.

Misra, G.K. and Gupta, R. (1981) *Resettlement Policies in Delhi*, Indian Institute of Public Administration: New Delhi.

Misra, H.N. (1994) 'Housing and Environment in an Indian City: A Case of a Squatter Settlement in Allahabad, India' in Hamish Main and Stephen Wyn Williams (eds.) *Environment and Housing in Third World Cities*, John Wiley and Sons: London.

Misra, S.N. (1985) (ed.), *Urbanisation and Urban Development in Punjab*, Guru Ram Dass P.G. School of Planning, Guru Nanak Dev University Press: Amritsar.

Mitra, B.C. (1990) 'Land Supply for Low Income Housing in Delhi' in Paul Baross et all (eds.) *The Transformation of Land Supply Systems in Third World Cities*, Avebury, Aldershot.

Mitra, B.C. and P. Nientied (1989) *Land Supply and Housing Expenses for Low Income Families: A Rationale for Government Intervention*, Urban Research Working Papers, No. 19, Free University, Netherlands.

Mohan, R. (1992) 'Housing and Urban Development Policy Issues for 1990's' *Economic and Political Weekly*, Vol. 27, No. 37, pp. 1990-1996.

Mohanty, B. (1993) *Urbanization in Developing Countries: Basic Services and Community Participation*, Concept: New Delhi.

Moser, C. (1982) 'A Home of One's Own: Squatter Housing Strategies in Guayaquil, Ecuador' in A. Gilbert et al (eds.) *Urbanization in Contemporary Latin America*, John Wiley, London, pp. 159-190.

Moser, C. (1987) 'Mobilisation is Women's Work: Struggle for Infrastructure in Guayaquil, Ecuador' in Moser, C. and Peake, L. (1987) (eds.) *Women, Human Settlements and Housing*, Tavistock Publishers: London.

Moser, C. (1989) 'Community Participation in Urban Projects in the Third World' *Progress in Planning*, Vol. 32, Part 2, Pergamon Press: UK.

Moser, C. (1992) 'Women and Self-Help Housing Projects: A Conceptual Framework for Analysis and Policy Making' in Mathey, K. (1992) (ed.) *Beyond Self-Help Housing*, Mansell: London.

Moser, C. (1993) *Gender Planning and Development: Theory, Practice and Training*, Routledge: London.

Moser, C. and Peake, L. (1987) (eds.) *Women, Human Settlements and Housing*, Tavistock Publishers: London.

Mukherjee, A. and Agnihotri, V.K. (1993) *Environment and Development: Views from the East and the West*, Concept: New Delhi.

Murison, H. and Lea, J.P. (1979) (eds.) *Housing in Third World Countries: Perspectives on Policy and Practice*, Macmillan: London.

Murthy, M.N. and Roy, A.S. (1993) 'The Development of the Sample Design of the Indian National Sample Survey during its First 25 Rounds' in M. Bulmer and D. Warwick (eds.) *Social Research in Developing Countries: Surveys and Censuses in the Third World*, UCL Press: London.

Nagpaul, H. (1996) *Modernization and Urbanization in India: Problems and Issues*, Rayat: New Delhi.

National Institute of Urban Affairs (1993) *Handbook of Urban Statistics*, New Delhi.

National Institute of Urban Affairs (1989) *Profile of the Urban Poor: An Investigation into their Demographic, Economic and Shelter Characteristics*, Research Study Series, No. 40.

Nelson, J.M. (1979) *Access to Power, Politics and the Urban Poor in Developing Nations*, Princeton University Press: Princeton.

Nientied, P., Meijer, E., and van der Linden, J. (1982) *Karachi Squatter Settlement Upgrading: Improvement and Displacement?*, Vrije Universiteit: Amsterdam.

Oberai, A.S. and Singh, H.K. (1983) *Causes and Consequences of Internal Migration: A Study in the Indian Punjab*, Oxford University Press: Delhi.

Oberoi, H.S. (1987) 'From Punjab to Khalistan: Territoriality and Metacommentary' *Pacific Affairs*, Vol. 60, No. 1, pp. 26-41.

Oberoi, H.S. (1994) *The Construction of Religious Boundaries: Culture, Identity and Diversity in the Sikh Tradition*, Oxford University Press: Delhi.

Palmer, E.K. and Patton, C.W. (1988) 'Evolution of Third World Shelter Policies' in Patton, Carl V. (1988) (ed.) *Spontaneous Shelter: International Perspectives and Prospects*, Temple University Press: Philadelphia.

Patton, C.V. (1988) (ed.) *Spontaneous Shelter: International Perspectives and Prospects*, Temple University Press: Philadelphia.

Patton, C.V. and Subanu, L.P. (1988) 'Meeting Shelter Needs in Indonesia' in Patton, C. V. (1988) (ed.) *Spontaneous Shelter: International Perspectives and Prospects*, Temple University Press: Philadelphia.

Payne, G. (1977) *Urban Housing in the Third World*, Leonard Hill: London.

Payne, G. (1984) (ed.) *Low-Income Housing in the Third World*, Wiley: London.

Payne, G. (1989) 'Informal Housing and Land Sub-Divisions in Third World Cities: A Review of the Literature' Prepared for the Overseas Development Administration (ODA), Centre for Development and Environmental Planning (CENDEP), Oxford Brookes University.

Peattie, L.R. (1979) 'Housing Policies in Developing Countries: Two Puzzles' *World Development*, Vol. 7, pp. 1017-22.

Peattie, L.R. (1982) 'Settlement Ugrading: Planning and Squatter Settlements in Bogota, Colombia' *Journal of Planning Education and Research*, Vol. 2, No. 1, 27-36.

Perlman, J. (1976) *The Myth of Marginality: Urban Poverty and Politics in Rio de Janeiro*, University of California Press.

Pettigrew, J. (1996) 'The State and Local Groupings in the Sikh Rural Areas, Post-1984' in Singh and Talbot (eds.) *Punjabi Identity: Continuity and*

Change, Manohar: New Delhi.

Pezzoli, K. (1995) 'Mexico's Informal Settlements' in Aldrich and Sandhu (eds.) *Housing for the Urban Poor: Policy and Practice*, Zed: New Delhi.

Pradilla, E. (1976) Noter om Boligproblemet [Notes on the Housing Problem] in Barnow [Translated from: Notas Acerca del Problema de la Vivienda, *Ideologia y Sociedad*, No. 16.]

Pugh, C. (1980) *Housing in Capitalist Societies*, Avebury: Aldershot.

Pugh, C. (1990) *Housing and Urbanisation: A Study of India*, Sage: New Delhi.

Pugh, C. (1995) 'The Role of the World Bank in Housing' in *Housing the Urban Poor: Policy and Practice in Developing Countries*, Zed: London.

Purewal, N.K. (1997) 'Displaced Communities: Some Impacts of Partition on Poor Communities' *International Journal of Punjab Studies*, Vol. 4, No. 1, Sage: New Delhi.

Qadeer, M.A. (1983) *Lahore: Urban Development in the Third World*, Vanguard Books: Lahore.

Ragin, C.C. (1994) *Constructing Social Research*, Pine Forge Press: Thousand Oaks.

Rai, S.M. (1986) *Punjab Since Partition*, Durga Publications: Delhi.

Raj, M. (1990) 'Zero cost administrative intervention in urban land market' in Baross and van der Linden (eds.) *The Transformation of Land Supply Systems in Third World Cities*, Avebury: Aldershot.

Rakodi, C. (1991) 'Developing Institutional Capacity to Meet the Housing Needs of the Urban Poor: Experience in Kenya, Tanzania and Zambia,' *Cities*, pp. 228-243.

Rakodi, C. (1995) *Harare: Inheriting a Settler-Colonial City: Change or Continuity?* John Wiley and Sons: London.

Ramachandran, R. (1991) *Urbanization and Urban Systems in India*, Oxford: Delhi.

Ramirez, R., Fiori, J., Harms, H. and Mathey, K. (1992) 'The Commodification of Self-Help Housing and State Intervention: Household Experiences in the Barrios of Caracas' in Mathey, K. (ed.) *Beyond Self-Help Housing*, Mansell: London.

Rao, P.S.N. (1991) 'Potentials, Limitations and Constraints of the Private Developer Housing Sector in India' paper presented at the International Housing Research Conference 'Housing Strategies for the 90's' Chandigarh, India, Sept. 25-28 1991.

Rodell, M.J. and Skinner, R.J. (1983) 'Introduction' in Skinner, R.J. and Rodell, M.J. (eds.) *People, Poverty and Shelter: Problems in Self-Help Housing in the Third World*, Methuen: London.

Rodney, W. (1976) *How Europe Underdeveloped Africa*, Bogle-L'Ouverture: London.

Rondinelli, D. (1983) *Secondary Cities in Developing Countries: Policies for Diffusing Urbanization*, Sage: Beverly Hills.

Rondinelli, D.A (1988) 'Increasing the Access of the Poor to Urban Services:

Problems, Policy Alternatives and Organisational Choices' in Rondinelli, D.A. and Cheema, G.S. (eds.) *Urban Services in Developing Countries: Public and Private Roles in Urban Development*, Macmillan: London.

Rondinelli, D.A. and Cheema, G.S. (1988) (eds.), *Urban Services in Developing Countries: Public and Private Roles in Urban Development*, Macmillan: London.

Roy, D.K. (1983) 'The Supply of Land for the Slums of Calcutta' in S. Angel et al (eds.) *Land for Housing the Poor*, Select Books: Singapore.

Saberwal. S. (1990) *Mobile Men: Limits to Social Change in Urban Punjab*, Manohar Publications: New Delhi.

Sandhu, R.S. (1987) 'Not All Slums are Alike' *Environment and Behaviour*, Vol. 19, No. 3, May, pp. 398-406.

Sandhu, R.S. (1988) *Evaluation & Impact of Slum Improvement Programme in Ludhiana (Punjab)*, Research Report 15, Indian Human Settlements Programme: New Delhi.

Sandhu, R.S. (1989) *The City and its Slums: A Sociological Study*, Guru Nanak Dev University: Amritsar.

Sarin, M. (1982) *Urban Planning in the Third World: The Chandigarh Experience*, Mansell Publishing Ltd.: London.

Sarin, M. (1983) 'The Rich, the Poor and the Land Question' in S. Angel et al (eds.) *Land for Housing the Poor*, Select Books, Singapore.

Schoorl, J.W. et al (eds.) (1983) *Between Basti Dwellers and Bureaucrats: Lessons in Squatter Settlement Upgrading in Karachi*, Pergamon: Oxford.

Selier, F. (1991) 'Family and Rural-Urban Migration in Pakistan: The Case of Karachi' in van der Linden and Selier (eds.) *Karachi: Migrants, Housing and Housing Policy*, Vanguard: Lahore.

Seliger, H.W. and Shohamy, E. (1989) *Second Language Research Methods*, Oxford University Press: Oxford.

Shahrashoub, R. (1992) 'Fieldwork in a familiar setting: the role of politics at the national community and household levels' in Devereux, S. and Hoddinott, J. (eds.) *Fieldwork in Developing Countries*, Harvester Wheatsheaf: London.

Shakur, T. (1988) 'Implications for Policy Formulation Towards Sheltering the Homeless: A Case Study of Squatters in Dhaka, Bangladesh' in *Habitat International*, Vol. 12, No. 2, pp. 53-66.

Shakur, T. (1992) 'Living Environment and Development: Approaches Towards the Improvement of 'Wet Components' in Low-Income Communities in the Third World: A Literature Review' proceedings from the 'Development and Planning Workshop' at SOAS, University of London: London.

Shakur, T. (1994) *Regulating the Un-Regulated Urban Housing Sub-Markets: An Antidote to Squatter Settlements?* Working Paper, Development and Environmental Studies Series, ICDES, Edge Hill College: Lancashire.

Sharma, H.C. (1996) *Artisans of Punjab: A Study of Social Change in Historical Perspective 1849-1947*, Manohar: Delhi.

Sharma, K.L. (1994) *Social Stratification and Mobility*, Rawat Publications: New Delhi.

Siddiqui, T.A. and Khan, M.A. (1990) 'Land Supply to the Urban Poor: Hyderabad's Incremental Development Scheme' in Baross and van der Linden (eds.) *The Transformation of Land Supply Systems in Third World Cities*, Avebury: Aldershot, pp. 309-335.

Siddiqui, T.A. and Khan, M.A. (1994) 'The Incremental Development Scheme' *Third World Planning Review*, 16 (3), pp. 277-291.

Singh, A.L., Fazal, S., Azam, F. and Rahman, A. (1996) 'Income, Environment and Health: A Household Level Study of Aligarh City, India' in *Habitat International*, Vol. 20, No. 1 , pp. 77-91.

Singh, F. (1990) (ed.) *The City of Amritsar: An Introduction*, Publication Bureau, Punjabi University, Patiala.

Singh, G. and Talbot, I. (1996) (eds.) *Punjabi Identity: Continuity and Change*, Manohar.

Singh, H. (1985) 'Patterns and Trends of Urbanisation in Punjab' in S.N. Misra (1985) (ed.) *Urbanisation and Urban Development in Punjab*, Guru Ram Dass P.G. School of Planning, Guru Nanak Dev University Press: Amritsar.

Singh, H. (1994) 'Housing Land Markets in Ludhiana City: A Study of Their Structure, Conduct and Performance' *Spatio-Economic Development Record*, Vol. 1, No. 5, November-December.

Singh, H. (1993) 'Demand for Housing Land: A Study of Formal and Informal Markets' *Punjab School of Economics Economic Analyst*, Vol. XIV, Nos. 1 and 2, Guru Nanak Dev University: Amritsar.

Singh, K. (1966) *A History of the Sikhs, Vol. II: 1839-1964*, Princeton University Press: Princeton NJ.

Singh, K. (1989) *The Partition of the Punjab*, Publications Bureau, Punjabi University: Patiala.

Singh, K. and Steinberg, F. (1996) 'Integrated Urban Infrastructure Development in Asia' *Habitat International,* Conference Report from the International Seminar on Integrated Urban Infrastructure Development 1-4 February, Indian Human Settlements Programme: New Delhi.

Singh, K., Steinberg, F. and von Einsiedel, N. (1996) (eds.) *Integrated Urban Infrastructural Development in Asia*, Intermediate Technology: London.

Singh, M. (1997) 'Bonded Migrant Labour in Punjab Agriculture' *Economic and Political Weekly*, Vol. XXXII, No. 11, March 15-21, pp. 518-520.

Singh, P. and Thandi, S. (1996) (eds.) *Globalisation and the Region: Explorations in Punjabi Identity*, Association for Punjab Studies: Coventry.

Singh, P. and Thandi, S. (1999) (eds.) *Punjabi Identity in a Global Context*, Oxford University Press: Delhi.

Sinha, A. (1991) 'Participant Observation: A Study of State-Aided Self-Help Housing in Lucknow, India' in Tipple, G.A. and Willis, K.G. (eds.) *Housing the Poor in the Developing World: Methods of Analysis, Case Studies and Policies*, Routledge: London.

Sjoberg, G. (1960) *The Preindustrial City: Past and Present*, Collier-Macmillan: London.

Sjoberg, G. (1965) 'Cities in Developing and in Industrial Societies: A Cross-Cultural Analysis' in P.M. Harris and L.F. Schnore (eds.) *The Study of Urbanization*, John Wiley and Sons: New York.

Skinner, R.J. and Rodell, M.J. (1983) *People, Poverty and Shelter: Problems of Self-Help Housing in the Third World*, Methuen: London.

Srinivas, M.N. (1980) *India: Social Structure*, Hindustan Publishing Corporation: New Delhi.

Stokes, C. (1962) 'A Theory of Slums' *Land Economics*, 38 (5).

Strassman, W. (1982) *The Transformation of Urban Housing: The Experience of Upgrading in Cartagena*, Johns Hopkins University Press: Baltimore.

Strassman, W.P. (1984) 'The Timing of Urban Infrastructure and Housing Improvement by Owner Occupants' *World Development*, Vol. 12, No. 7, pp. 743-54.

Struyk, R.J. (1988) *Assessing Housing Needs and Policy Alternatives in Developing Countries*, The Urban Institute Press: Washington D.C.

Suttles, G.D. (1972) *The Social Construction of Communities*, University of Chicago Press: Chicago.

Tait, J. (1997) *From Self-Help to Sustainable Settlement: Capitalist Development and Urban Planning in Lusaka, Zambia*, Avebury: Aldershot.

Talbot, I. (1996) 'State, Society and Identity: The British Punjab' in Singh and Talbot (eds.) *Punjabi Identity: Continuity and Change*.

Tanphipat (1983) 'Immediate Measures for Increasing the Supply of Land for Low-Income Housing in Bangkok' in Angel, S. et al (eds.) *Land for Housing the Poor*, Select Books, Singapore.

Tatla, Darshan (1998) *The Sikh Diaspora: In Search of Statehood*, UCL Press: London.

Thandi, S.S. (1996) 'Counterinsurgency and Political Violence in Punjab' in Singh and Talbot (eds.) *Punjabi Identity: Continuity and Change*.

Thorbek, S. (1994) *Gender and Slum Culture in Urban Asia*, Zed: London.

Tipple, G., Amole, B., Korboe, D. and Onyeacholem, H. (1994) 'House and Dwelling, Family and Household: Towards Defining Housing Units in West African Cities' *Third World Planning Review*, Vol. 16, No. 4, pp. 427-450.

Todaro, M. (1969) 'A Model of Labour Migration and Urban Unemployment in Less Developed Countries' *American Economic Review*, No. 59.

Todaro, M. (1976) *Internal Migration in Developing Countries*, ILO: Geneva.

Todaro, M. (1994) *Economic Development*, Longman: London.

Tully, M. and Jacob, S. (1985) *Amritsar: Mrs. Gandhi's Last Battle*, Pan Books Ltd.: London.

Turner, J.F.C. (1963) 'Dwelling Resources in South America' *Architectural Design*, No. 37, pp. 507-26.

Turner, J.F.C. (1967) 'Barriers and Channels for Housing Development in Modernizing Countries' *Journal of the American Institute of Planners*,

Vol. 33, pp. 167-81.

Turner, J.F.C. (1968) 'Housing Priorities, Settlement Patterns and Urban Development in Modernizing Countries' *Journal of the American Institute of Planners*, 34, 354-63.

Turner, J.F.C. (1969) 'Uncontrolled Urban Settlements: Problems and Solutions' in Breese, G. (ed.) *The City in Developing Countries*, Prentice-Hall: Englewood Cliffs, NJ.

Turner, J.F.C. and Fichter, R. (1972) *Freedom to Build: Dweller Control of the Housing Process*, Macmillan: New York.

Turner, J.F.C. (1976) *Housing by People: Towards Autonomy in Building Environments*, Marion Boyars: London.

Turner, J.F.C. (1978) 'Housing in Three Dimension: Terms of Reference for the Housing Question Redefined' *World Development*, Vol. 6, No. 9/10, pp. 1135-45.

Turner, J.F.C. (1982) 'Issues in Self-Help and Self-Managed Housing' in Ward, P. (ed.) *Self-Help Housing: A Critique*, Mansell: London.

United Nations Centre for Human Settlements (HABITAT) (1996) *An Urbanizing World: Global Report on Human Settlements*, Oxford University Press: Oxford.

United Nations Development Programme (1996) *Human Development Report 1996*, Oxford University Press: New York.

United Nations Secretariat (1952) *Urban Land Policies*, Document No. ST/SCA/9, New York.

van der Harst, J. (1983) 'Financing Housing in the Slums of Karachi' in Schoorl et al (eds.) *Betweem Basti Dwellers and Bureaucrats*, Pergamon: Oxford.

van der Linden, J. (1983) 'The Bastis of Karachi: The Functioning of an Informal Housing System' in Schoorl, J.W., van der Linden, J. and Yap, K.S. (eds.) *Between Basti Dwellers and Bureaucrats: Lessons in Squatter Settlement Upgrading in Karachi*, Pergamon: Oxford.

van der Linden, J. (1983) 'The Squatter's House as a Source of Security' in Schoorl, J.W., van der Linden, J. and Yap, K.S. (eds.) *Between Basti Dwellers and Bureaucrats: Lessons in Squatter Settlement Upgrading in Karachi*, Pergamon: Oxford.

van der Linden, J. (1986) *The Sites and Services Approach Renewed*, Gower: Aldershot.

van der Linden, J., Nientied, P. and Kalim, S.I. (1990) 'Pakistan' in van Vliet, Willem (ed.) *International Handbook of Housing Policies and Practices*, Greenwood Press: New York.

van der Linden, J. (1992) 'Back to the Roots: Keys to Successful Implementation of Sites-and-Services' in Mathey, Kosta (ed.) *Beyond Self-Help Housing*, Mansell: London.

van Lindert, P. (1992) 'Social Mobility as a Vehicle for Housing Advancement?: Some Evidence from La Paz, Bolivia' in Mathey, K. (ed.) *Beyond Self-Help Housing*, Mansell: London.

van Westen, A.C.M. (1990) 'Land Supply for Low-Income Housing in Bamako,

Mali; Its Evolution and Performance' in Baross and van der Linden (eds.) *The Transformation of Land Supply Systems in Third World Cities*, Avebury: Aldershot.

Vanaik, A. (1990) *The Painful Transition: Bourgeois Democracy in India*, Verso: London.

Wahab, E.A. (1991) 'The Tenant Market of Baldia Township' in van der Linden and Selier (eds.) *Migrants, Housing and Housing Policy*, Vanguard: Lahore.

Wallerstein, I. (1979) *The Capitalist World Economy*, Cambridge University Press: Cambridge.

Ward, P. (1976) 'The Squatter Settlements as Slum or Housing Solution: Evidence from Mexico City' *Land Economics*, 52, pp. 330-46.

Ward, P. (ed.) (1982) *Self-Help Housing: A Critique*, Mansell Publishing Ltd.: Oxford.

Ward, P. (1983) 'Land for Housing the Poor: How can Planners Contribute' in Angel et al (eds.) *Land for Housing the Poor*.

Ward, P. and Chant, S. (1985) 'Community Leadership and Self-Help Housing,' *Progress in Planning*, Vol. 27, Part 2, pp. 69-136.

Ward, P. and Macaloo, C. (1992) 'Articulation Theory and Self-Help Housing Practice in the 1990's' *International Journal of Urban and Regional Research*, Vol. 6.

Wegelin, E.A. (1995) 'Squatter and Slum Settlements in Pakistan: Issues and Policy Challenges' in Aldrich, B. and Sandhu, R.S. (eds.) *Housing the Urban Poor: Policy and Practice*, Zed Books: New Delhi.

Wegelin, E.A. and Chanond, C. (1983) 'Home Improvement, Housing and Security of Tenure in Bangkok Slums' in Schoorl, van der Linden and Yap (eds.). *Between Basti Dwellers and Bureaucrats: Lessons in Squatter Settlement Upgrading in Karachi*, Pergamon: Oxford.

Whyte, W.F. (1964) 'High Level Man Power for Peru' in C. Myers and F.H. Harbison (eds.) *Manpower and Education*, Mc Graw-Hill: New York.

Whyte, W.F. and Alberti, G. (1976) *Power, Politics and Progress: Social Change in Rural Peru* Elsevier: New York.

Whyte, W.F. and Alberti, G. (1993) 'On Integrating Research Methods' in Bulmer, M. and Warwick, D.P. (eds.) *Social Research in Developing Countries: Surveys and Censuses in the Third World*, UCL Press: London.

Wiebe, P.D. (1975) *Social Life in an Indian Slum*, Vikas Publishing House: Delhi.

Wuelker, G. (1993) 'Questionnaires in Asia' in Bulmer and Warwick (eds.) (1993) *Social Research in Developing Countries*, UCL Press: London.

Yahya, S.S. (1990) 'Residential Urban Land Markets in Kenya' in P. Amis and P. Lloyd (eds.) *Housing Africa's Urban Poor*, Manchester University Press: Manchester.

Yap, K.S. (1996) 'Low-Income Housing in a Rapidly Expanding Urban Economy: Bangkok 1985-1994' *Third World Planning Review*, Vol 18, No. 3, August, pp. 307-324.

Yap, K.S. (1982) *Leases, Land and Local Leaders in Karachi: An Analysis of a*

Squatter Settlement Upgrading Programme in Karachi, Vrije Universiteit: Amsterdam.

Zetter, R. (1984) 'Land Issues in Low Income Housing' in G. Payne (ed.) *Low-Income Housing in the Third World*, Wiley: London.

Zorbaugh, H.W. (1929) *The Gold Coast and the Slum: A Sociological Study of Chicago's Near North*, University of Chicago Press: Chicago.

Index

T - #0544 - 101024 - C0 - 219/152/12 - PB - 9781138728714 - Gloss Lamination